How A Nuclear Incident In America Will Probably Go Down:

What We Learned From Three Mile Island, Chernobyl and Fukushima

Dr. Charles S. Brocato

Text and Images © 2018 Charles S. Brocato except where otherwise noted

All rights reserved

www.chemicalbiological.net

Cover photo: istock.com/vencavolrab

All Other Photos: Charles S. Brocato

Dedicated to

St. Martin de Porres

Who Always Answers Prayer

Acknowledgements

Many thanks to Kathryn E. King, who aided me with every aspect of this book. Without her encouragement and help, this project might never have been completed.

Table of Contents

Acknowledgements 4

Introduction 7

Part I: The Accidents 15

2: The Accident At Three Mile Island 17

3: The Accident At Chernobyl 33

4: The Accident At Fukushima 47

Part II: Lessons Learned 63

5: It's Worse Than You Think 65

6: Nobody Knows Anything 69

7: Operators Disbelieve Their Instruments 73

8: First Responders Are Ill-Equipped 77

9: Rumors Abound 83

10: The Big Question: Who's In Charge? 89

11: How The Next Event May Go Down 97

Radiation Limits For Outside Activities 117

Link To Downloads Page 118

About the Author 123

Other Books By This Author 125

Introduction

When Dr. "B" first began studying radiation, he studied the accident at Chernobyl with great interest. He learned a lot from it, including the fact that in a serious nuclear incident, you need a serious radiological instrument that will read high levels of radiation. Many of the meters being sold today are designed to read the very low levels in food, not the hundreds or thousands of Roentgens likely in a serious nuclear incident.

He also observed many other things as the accident at Chernobyl progressed and authorities attempted to cover up the incident, evacuate the area around the plant and bury the radioactive remains of the Chernobyl No. 4 reactor.

From there, he studied the accidents at Three Mile Island and at Fukushima. Although the causes of these three accidents differed, he observed many likenesses in the responses of the plant workers as they tried to understand what had happened and figure out what to do about it.

The responses of government when it got involved were equally fascinating. Even though these three accidents occurred in three different countries with very different types of government, the authorities displayed some similar responses as they tried to deal with the aftermath of the accidents and calm their populations.

According to Dr. "B," we can study these three accidents and gain a very good understanding of what the response of the United States government is likely to be like if a serious nuclear disaster happens in America, whether it's a nuclear plant accident or a nuclear bomb exploding somewhere in America...or maybe even a dirty bomb explosion.

The bigger the disaster, he says, the more you can look for the responses he will describe in this book.

Forewarned is forearmed. If you believe America is likely to suffer a serious nuclear event in the near future, no matter what the source, this book will give you a good idea of what to expect.

1: What Can We Learn From Past Accidents?

Those who cannot remember the past are doomed to repeat it, or so it is said. A truer statement is probably along these lines: Those who remember the past are likely to repeat the same mistakes made in the past because this time it's different.

We are not sure why this is...or maybe we are. It's because the mistakes in these situations are usually made by governments, and governments are the same everywhere, no matter what their ideologies.

If a nuclear incident occurs in America, whether it's a nuclear bomb exploding in a city, a nuclear power plant meltdown, or some awful occurrence at one of the nation's nuclear sites, you can bet on one thing. America will be temporarily paralyzed.

As happened on September 11, 2001, when terrorists destroyed the World Trade Center in New York City, people all over America were in disbelief and horror. We suddenly remembered what churches were for, and Congress got busy trying to look as it was doing something and passed the Patriot Act.

When the anthrax-by-mail attacks happened in the fall of 2001, people all over America, including those in small towns far removed from the post offices that actually handled the "white-powder" letters, approached their mailboxes in deadly fear. Many people panicked at the very thought of dying with anthrax, and some even experienced symptoms of what they thought might be anthrax.

If a nuclear attack or a bad nuclear accident happens in America, look for this to repeat.

People located hundreds of miles from the center of the attack are likely to experience symptoms of ARS (Acute Radiation Syndrome). They will monitor the skies for signs of deadly fallout, the rivers and streams for radioactive particulates and the news for hints of where the next attack or event may take place.

In short, we will suddenly find ourselves living in "interesting times."

We began studying accounts of the nuclear accidents of the past because we found them so interesting, especially the accident at Chernobyl. As noted earlier, we learned that in any serious nuclear incident, you will need a radiological instrument capable of reading high levels of radiation.

Because of the current idea that any amount of radiation is dangerous, most radiological meters sold these days measure only very low doses of radiation. They would not be of much use in a real nuclear fallout situation unless the fallout was just beginning to arrive or had decayed to very low "safe" levels.

We considered that realization a valuable lesson and began studying other books on famous nuclear accidents in search of other such lessons.

Fukushima had just happened, and we followed it avidly. What became very clear to us was that the Japanese people mistrusted their government's figures and relied on their own radiation meters and their own wind-direction calculations to learn where any radioactive plumes might travel. Those in the "know" were able to avoid areas, foods and situations that might have affected their health.

In fact, radiation meters sold out in Japan. Many Japanese citizens ordered meters online from American companies. People were seen in supermarkets testing their food with meters before buying.

Nuclear Hysteria In America

We also saw that nuclear hysteria held America captive. You could not buy a bottle of potassium iodide tablets anywhere, even though levels of radioactive iodine were extremely low in America and not likely to rise much higher. The fact that one or two atoms of iodine-131 had been detected in American milk was enough to sell thousands of bottles of KI.

In California, people with low-level meters measured the radiation given off by the beach sands and clicked their tongues in disapproval when the meter buzzed away and gave them a positive measurement. Never mind that the reading was the normal background level of radiation found in that area. After all, radiation is radiation, and it's all bad.

Except that it isn't. Every area of the United States of America has a *background level* of radiation. This is the natural level that comes from radioactive elements present in the soil and rocks in the area (or the beach sands), and from cosmic radiation that comes in from outer space. If you do not know the normal background radiation level of the area you measure, you will not know if a low-level reading indicates the presence of outside radiation, or if it's the normal level. For this reason, if you have (or get) a radiological meter, be sure to take background readings in your home and around your property and write them down for future reference.

We saw a lot of new meters in use during the early stages of the Fukushima accident. People would get a brand new low-level meter and start taking readings, firmly believing that if they got any reading at all, then it had to indicate the presence of Fukushima radiation. They had no idea there was such a thing as background radiation.

Some areas of the U.S., such as Colorado, boast a much higher level of natural background radiation than other areas because of a higher level of radioactive elements in the surrounding rocks. We have no doubt that people with new meters who took readings in those areas thought for sure Fukushima radiation had landed right there.

Normalcy Bias

It is likely that much of the response to all three of these major accidents can be explained by what's known as *normalcy bias*. This is a mindset, one we all tend to have, that causes people to underestimate the possibility of a disaster or the possible effects of a disaster. After all, disasters don't happen every day. If they did, we'd expect them.

This phenomenon is what causes people to sit home when the television weather guy announces that a hurricane will be hitting their area in less than 24 hours. These are the people who shrug off the warning and note that TV weather announcers have been predicting storms for years and not one ever showed up. Then they wander around their darkened neighborhoods after the storm trying to borrow batteries for their flashlights and a bottle or two of water until FEMA can arrive in a few more days with the MREs and the cases of water.

This bias causes all of us to assume that tomorrow is going to be like today, and nothing untoward is going to happen. Until it does, and we are left stunned and wondering how on earth such a thing could have happened. In fact, maybe we are dreaming and it didn't happen after all.

In nuclear power plants, this bias, which is also known as *analysis paralysis*, or *the ostrich effect*, usually results in inadequate disaster planning. No one has thought out what could happen or made any contingency plans, and no evacuation plans are made because everyone knows that a meltdown can never occur. There are just too many safety precautions to prevent that sort of thing.

As in Nazi Germany, where many wealthy and prosperous Jews remained in Germany as the signs mounted that they would be wise to flee the country while they still could, the workers at the three nuclear power plants we are considering were more or less blindsided by the accidents. When their control panel readings clearly showed things were not as they should be, they were baffled. Something had to be wrong with the readings!

At Three Mile Island, workers could not figure out what was causing the radiation alarms to go off. This went on for hours until a new crew came on and discovered that the pilot relief valve was stuck open, even though the control panel showed it to be closed.

At Chernobyl, workers in the control room finally realized the reactor was unstable, but by then it was too late. When the unthinkable happened and the reactor exploded due to a buildup of steam, they clung to the belief that the reactor was still intact and just needed some cooling water.

At Fukushima, nobody had ever thought that several reactors could melt down at once, or that a tsunami high enough to top the protective sea wall and drown the emergency generators could occur. The emergency generators kept cooling water circulating through the reactor cores when the power failed, and once those generators died, the plant lost all means to keep the reactor cores cool.

Often, we tend to think that if we don't plan for these things, they can't happen. For instance, many people refuse to write a will on the grounds that if they do, somehow or other this will cause them to die. If you don't write a will, then you can't die.

Studying The Accidents

As they say, history tends to repeat, and if it doesn't repeat in exactly the same way, then it definitely rhymes.

We learned so much from studying Chernobyl that we decided to study the accidents at Three Mile Island and Fukushima. Sure enough, all three of those accidents shared certain features that tell us a lot about how a nuclear event in America will likely go down.

So let us look into each accident and take note of what happened, both at the plant and with the people of the country.

You will find them quite enlightening.

Part I: The Accidents

2: *The Accident At Three Mile Island*

March 28, 1979

Three Mile Island is an island located in the Susquehanna River about three miles downriver from Middletown, Pennsylvania. A nuclear power plant, the Three Mile Island Nuclear Generating Station, was built there that operated two reactors.

The famous accident that happened there is still rated as the most significant nuclear incident in American power plant history and is rated a 5 on the seven-point scale of International Nuclear Events, as it was considered an "accident with wider consequences."

The accident itself was a classical *loss-of-coolant accident*, wherein the core of the reactor somehow loses its flow of cooling water. The reactor core then overheats, as the nuclear chain reaction generates lots of heat that must be dissipated, or the excessive heat damages the nuclear fuel rods. Even when the reactor has been shut down, the fuel keeps generating *decay heat*, which must also be dissipated.

When the nuclear fuel rods lose their zirconium cladding to the excessive heat, the fuel rods begin to warp and melt. To add to the trouble, the zirconium cladding reacts chemically with water or steam to release hydrogen gas.

The accident at Three Mile Island's number 2 reactor began when workers attempted to clear a blockage in one of the reactor's condensate polishers, a filter responsible for cleaning minerals and other impurities from the water in order to prevent corrosion in the steam generators. The operation did not go as usual and caused the feedwater pumps and other pumps to turn off at around 4 a.m., which in turn caused a turbine to trip.

Once this happened, the reactor's instruments noted a lack of water and performed an emergency shutdown, known as a SCRAM. Control rods were automatically inserted into the core and the chain reaction was halted, but the reactor continued to generate decay heat.

The main factor causing the accident was the fact that when the feedwater pumps ceased operation, three auxiliary pumps came on automatically. However, the valves had been closed for routine maintenance and no water could be pumped to the core. These valves were not supposed to be closed, and if they had to be closed for maintenance, the reactor was supposed to be shut down. This was considered a major error on the part of the operators.

On top of that, a key valve, the pilot-operated relief valve that was intended to relieve excess pressure, suffered a mechanical fault and got stuck in the open position, which allowed cooling water to escape from the system. This failure is considered the primary cause of the loss of coolant and the partial core meltdown that followed.

As if to make everything worse, the valve was stuck open but the indicator light on the control panel in the control room indicated the valve was closed. In actuality, the light did not indicate the position of the valve at all, but that the solenoid powering it was receiving power or not receiving power. Until an operator actually went and looked at the valve some hours later, the reactor problem could not be diagnosed.

This situation is considered another major flaw of the system. The indicator light did not actually indicate the real position of the valve unless everything was operating correctly. The operators were used to everything operating correctly and did not have any idea that the valve could actually be open when the indicator light said it was closed.

In addition, a temperature indicator that might have suggested that the valve was actually open was located in a position that was out of the operators' sight.

Another indicator that unfortunately did not exist at the time was an indicator of the actual water level in the core. Operators used the water level in another unit called the pressurizer unit instead, assuming that if it was high, then the water must be covering the core. In actuality, steam had begun to fill the water pipes and create "steam voids." This misreading of the pressurizer water level was one of the main reasons why no one realized the core had actually lost its cooling water and was dangerously overheating.

By the time a new shift came on duty and one of the new arrivals noted the high temperature on the indicator that was out of everyone's sight, radioactive water from the core had already flowed into the containment building and seriously contaminated it.

How The Accident Went Down

At 4:00 a.m. on Wednesday, March 28, the feedwater pumps at the Three Mile Island No. 2 reactor suddenly shut down, and the auxiliary feed pumps came on. But with their valves closed, the auxiliary pumps could not pump anything. The heat buildup caused pressure in the primary loop to increase, which the pilot relief valve was supposed to relieve by opening. Unfortunately, the valve stuck in the open position and cooling water flowed out.

Operators Disbelieve Their Own Instruments

At 4:15 a.m., radioactive coolant leaked into the general containment building and the containment building sump pumped the contaminated water into an auxiliary building. This caused a radiation alarm to sound. However, the operators ignored this in the belief that the core was safely covered with water.

Soon, alarms were sounding all over the control room, and nobody could figure out why. Some instrument dials reported readings that were well outside the norm, but the workers remained baffled. They had not been well-trained in trouble scenarios.

Basically, the operators ignored the alarms and the readings their instruments were giving them, firm in the belief that the pilot relief valve was closed and the core was properly covered with cooling water. Thus, for more than two hours, the core was losing coolant and overheating while the workers sought for other causes of the problems.

Not even the Nuclear Regulatory Commission (NRC) staff who had arrived onsite believed the instruments. In fact, one NRC staff member told the NRC commissioners that although "fuel failure" had occurred, they didn't know how much had melted down. In *Three Mile Island: A Nuclear Crisis in Historical Perspective*, J. Samuel Walker observes, "He (the NRC staff member) also reported that radiation levels in the containment dome were measuring 20,000 rads per hour, but submitted that this extraordinarily elevated reading was probably the result of 'an instrument problem.'"

It's Worse Than You Think

When an accident begins, nobody wants to believe what's happening is actually something bad. The operators usually figure that if they can just hit the right button, the problem will resolve itself. When that doesn't happen right away, nobody is quite sure what to do next.

At 6:57 a.m., a supervisor declared a *Site Area Emergency*. According to the U.S. Nuclear Regulatory Commission (NRC), this means "events are in progress or have occurred which involve actual or likely failures of plant functions needed for protection of the public," and that "any releases are not expected to result in exposure levels which exceed EPA PAG (Environmental Protection Agency Protective Action Guide) exposure levels beyond the site boundary."

The purpose of this level of alert is to make sure all personnel are at their posts, that monitoring teams are ready to go, and that "personnel required for evacuation of near-site areas are at duty stations if the situation becomes more serious, to provide consultation with offsite authorities, and to provide updates to the public through government authorities." Any releases of radioactive material would be within the boundaries of the plant.

General Emergency

Less than half an hour later, the station manager announced a *General Emergency*, which indicates that "events are in progress or have occurred which involve actual or imminent substantial core degradation or melting with potential for loss of containment integrity," and "releases can be reasonably expected to exceed EPA PAG exposure levels offsite for more than the immediate site area."

This means that a serious accident is in progress that may lead to a core meltdown, and that releases of radioactive isotopes of higher levels can occur, and may extend beyond the boundaries of the plant.

The purpose of this declaration is to "initiate predetermined protective actions for the public," not to mention providing a "continuous assessment of information" from the licensee (the nuclear plant) and offsite organizational measurements," and to prepare to carry out whatever actions are deemed necessary to deal with any actual or potential releases of nuclear materials.

Nobody Knows What's Going On

Once a General Emergency is declared, information is supposed to start flowing to everyone who needs to know, and to the public. Everyone is fully informed, and everyone knows what to do.

Or perhaps not.

The Three Mile Island incident is a great example of what actually happens.

Metropolitan Edison (Met Ed) the owner of the plant, notified the Pennsylvania Emergency Management Agency, which then notified state and local agencies, the governor, and the lieutenant governor, as was proper.

From there, everything went downhill. Since the operators working at the consoles didn't know what was going on, neither did the supervisors or the plant manager. So what they told Met Ed was repeated to the press and to state government officials, but with a bias toward "no radiation releases."

The governor held a press conference and said there had been a "small" radiation release as he had been told by his informant. He was promptly contradicted by Met Ed and by another official, both claiming that no radioactive particulates had been released. The governor was not happy. He became even less happy when the plant vented steam without notifying him, as they should have.

State officials, convinced that Met Ed was understating the severity of the accident and angry that the plant had conducted a steam venting without informing them in advance, contacted the NRC.

The NRC, having also been alerted by Met Ed about the accident, sent staff members to Three Mile Island and activated its emergency response center in Bethesda, Maryland. Not that it did much good. The NRC was not really equipped to deal with real emergencies and had no well-laid-out command structure, nor did the agency have any authority to give orders to Met Ed or anyone else.

On top of that, the NRC had the same trouble everyone else was having in trying to find out what was actually going on.

One of the NRC staff members sent to Three Mile Island wrote years later that the first thing he observed was the lack of a clear and well-defined organizational structure. The plant manager was supposedly in charge overall, but when problems arose, everyone got together to brainstorm the best idea to deal with it. The confabs even included the men from the NRC, in case they should have some better ideas about how to solve the problem.

The truth is, much of what actually happened in the Three Mile Island incident wasn't known until years later. Or, if anyone knew it earlier, it was never told to any of the concerned officials, much less to the public. According to one of the NRC officials, it took 5 weeks for him to learn that operators had actually measured temperatures of the fuel that were almost at the melting point.

Nobody knew a large part of the core had melted down until cleanup workers actually opened the reactor vessel several years later. If this had been known—and many say it *should* have been known from all the indications present—it is doubtful that the information would have been made public.

The Press Versus A Morass of Conflicting Info

In the meantime, the news media had begun arriving to cover what looked like a small nuclear plant accident, just in case it amounted to anything. As the incident developed, every major news outlet in the country sent reporters, and some of these reporters were very experienced in getting needed information.

In this case, however, reporters seemed to be getting different stories, and what one reported usually contradicted what another reporter wrote or said. Needless to say, members of the press became really angry and some reporters actually refused to report what certain individuals said, such as the executives from Met Ed, whom they regarded as liars almost from the very first.

Everyone, including the public, longed for one trustworthy voice that would tell them the truth about what was going on so the reporters could report truthful information and so the public could figure out what to do. Many of the designated spokespersons did not know the answers to questions, as they were spokespeople, *not technical people*, and others contradicted previous sources.

After the incident had progressed and confusion continued to build, President Jimmy Carter sent an NRC representative named Harold Denton to the site to serve as his personal representative. Denton turned out to be someone the press and the people of Pennsylvania could trust as well. As one reporter put it, he looked like a regular guy and had a manner that inspired trust.

Three days into the incident, operators discovered a big hydrogen bubble had developed in the dome of the pressure vessel. If it exploded, it could burst open the containment vessel and release a large amount of radioactive material. The press had plenty to report, what with speculations about whether or not the bubble would explode, until scientists finally determined that it would not explode since there was no oxygen present.

The discovery of the hydrogen bubble created a huge furor and concern among the press and the public. Nobody could find out enough information to know whether or not they should evacuate. The authorities in charge of the evacuation knew just about as much as the public and worried about whether or not they would be able to get any evacuation started soon enough to clear out the area before the bubble exploded. The governor and the emergency operations director had the very same questions and concerns.

Rumors Abound

In the meanwhile, somebody claimed an evacuation had been ordered, which was immediately countered by other officials. The infighting among government agencies represented at Three Mile Island was intense by itself, and nobody ever took responsibility for the evacuation order. Indeed, many people had already fled the area on their own, figuring they might as well get ahead of the radiation they were sure was coming.

In addition to this, rumors rolled back and forth about whether an evacuation would be called and whether or not a radiation release of varying sizes had already occurred or was about to occur.

One rumor held that Middletown, Pennsylvania, had been "completely obliterated." Another rumor said that the entire population of Middletown had been evacuated by helicopter.

People rushed to their banks and withdrew cash, so much so that the Federal Reserve sent more cash to the area's banks. Some feared the cash would become radioactive if they left it in the bank, an illustration of how little people know about radioactivity.

These intense concerns lasted for about a day, until workers instituted procedures to shrink the hydrogen bubble. Within a couple of days that danger was past.

During all of the rush to flee and the ongoing press coverage, nobody was quite sure whether or not there had been, or would be, a release of radioactive isotopes into the environment.

As one of the NRC staffers wrote in 2009, nobody learned for years, until the reactor vessel was actually opened, that by the time the NRC was called in, about 8 a.m. the morning the accident began, that about half of the uranium fuel rods had already melted.

The Radioactive Release

Actually, there had been a release of radioactive material, back on the first day of the incident, while the pilot relief valve was stuck open and nobody knew it. A lot of coolant, namely water contaminated with radioactive products of fission, flowed out through the valve and wound up in the auxiliary building, which was outside the containment boundary.

Radiation alarms began to sound, and the auxiliary building was certainly contaminated, but very little radioactivity escaped into the atmosphere. This was because most of what entered the atmosphere was in the form of the noble gases xenon and krypton, which are inert (chemically unreactive) even though they are radioactive. Some radioactive iodine was released also, as it is a gas, but the amounts were very low, so low elevated levels were never found in milk samples.

The Environmental Protection Agency (EPA) began monitoring the environment within hours after the accident began, sampling the air, water, soil, sediments and plants. They never found any contamination to speak of.

Researchers with extremely sensitive equipment that could detect Chinese atmospheric nuclear testing found no elevated levels of radioactivity in the entire area.

A Best-Case Accident

Three Mile Island could have ended in a serious disaster, but in spite of the dire possibilities, the reactor was shut down and the only major loss was the reactor itself. In the end, the Three Mile Island incident was rated a 5 on the 7-point International Nuclear Event Scale as an "Accident with wider consequences."

Accidents with a rating of 5 usually involve severe damage to the reactor core and a release of large amounts of radioactive material within an installation, with the possibility of significant public exposure. An unknown amount of radioactive gases were released into the atmosphere, but no deaths can definitively be attributed to the incident.

Because of this accident, groups protesting nuclear power became much more credible. This was the first American accident to receive such extensive press coverage, so much so that Americans became aware of exactly what can go wrong with a nuclear power plant. Before, most people had believed the nuclear industry when it claimed nuclear power was clean, safe and cheap.

Thanks to the Three Mile Island accident, nuclear plant control rooms were redesigned, so that pertinent gauges were in the sight of operators and indicated the true state of the valve or machine they monitored. Nuclear plants were required to write up evacuation plans for the public living within ten miles of the plant.

Before this accident, there had been no preparation of the public for a nuclear accident at a power plant because, after all, nuclear power was so safe, a serious accident could not happen.

Operators Ill-Equipped To Deal With The Disaster

Studies of the accident to determine the causes settled upon this belief as being one of the main drivers behind the accident: *Nuclear power in the United States is so safe, there is no need to consider extra precautions.*

The Three Mile Island workers and everyone concerned believed that nuclear power was already so safe and had so many redundant safety precautions built in, there was no need to consider any of the outside-the-ordinary things that could possibly go wrong.

In other words, *people* were found to be the fundamental problem that caused the Three Mile Island accident.

It's that same old problem again: If you think about something for purposes of advance preparation, you're going to bring it on, so don't think about it.

Compounding The Problem

Not even two weeks before the accident, the movie, *The China Syndrome*, starring Jane Fonda, Michael Douglas and Jack Lemmon, premiered. At first, the nuclear industry protested loudly against the fictionalized account of a nuclear plant supervisor who discovers serious safety violations at his plant and seeks to publicize them.

The name comes from the idea that if a nuclear reactor melts down, the radioactive core will melt all the way through the reactor containment and head through the earth all the way to China, contaminating ground water in the area and sending up radioactive steam.

The reporter, the camera operator and the nuclear plant supervisor join forces to expose the serious problem at the nuclear plant, opposed all the way by the plant's owners. Sure enough, the feared accident begins at the end of the movie, resulting in the death of the supervisor and the reporter getting her story of the exceedingly narrow escape from a horrendous accident as the reactor is brought back under control.

The movie raised a lot of questions about the true safety of nuclear power plants at a really bad time for the nuclear industry.

Who's In Charge?

Any time several government agencies are involved in an incident, each of those agencies wants priority and believes it is the agency in charge.

In the Three Mile Island incident, the Environmental Protection Agency vied with the Nuclear Regulatory Commission and the Department of Energy for top dog, with the top honors finally winding up with Harold Denton of the Nuclear Regulatory Commission.

At least, he was the only person involved at the top who seemed to really know what was going on and could tell it to other people in terms they could understand.

According to J. Samuel Walker, in *Three Mile Island: A Nuclear Crisis in Historical Perspective*, the NRC's response to the hydrogen-bubble crisis "vividly demonstrated its lack of command structure and its weakness as an operational agency. The roles and authority of the commission and the staff were ill defined. One result is that senior staff members in the incident response center recommended an evacuation to the state based on erroneous information, without consulting the agency's personnel at the site or the commission." Fortunately, state officials and a few of the NRC staff members on site knew more about the crisis and were able to stop the evacuation announcement.

Who's in charge? is one of the biggest problems we see in most accidents. At Three Mile Island, nobody was quite sure at first who was in charge because nobody knew for sure what was happening at the plant, including the operators, the supervisors and the owners.

Years later, this same question was the chief item learned from joint government-run operations to test preparedness for terrorism events and other disasters. The various government agencies involved did not know who the Number One Head Honcho was, and each agency usually thought it was the premier solver of all problems.

Providing an answer to this question is why federal agencies involved in emergency management created the Incident Command System (ICS). This alone resolves an enormous amount of confusion as the response effort comes together.

As you may have observed, several other signposts occurred in the Three Mile Island accident, and these can usually be seen in other accidents, including the next two we will consider, Chernobyl and Fukushima Daiichi.

Look for these signposts as you read the next two chapters, not necessarily in this order:

> Who's in charge?
>
> It's worse than you think.
>
> Nobody knows what's going on.
>
> Operators disbelieve their own instruments.
>
> First responders and operators are ill-equipped to deal with the disaster.
>
> Rumors abound.

Every accident showcases several to all of these points. They are responsible for much of the confusion that accompanies major disasters.

3: The Accident At Chernobyl

April 26, 1986

The Chernobyl accident, usually known as the Chernobyl disaster, began as a power surge that was caused by powering down the reactor for a turbine test. The reactor became unstable and overheated, the cooling water flashed to steam, and a massive steam explosion followed. A second explosion followed that blasted a tremendous amount of radioactive core material into the skies, scattering radioactive debris all over the plant site and sending huge radioactive plumes over various European countries.

Soon after midnight, on the morning of April 26, in the Soviet state of Ukraine, operators of the Chernobyl number 4 reactor began running a turbine test to determine whether or not a reactor could be cooled by the coasting-down turbines if a power outage occurred. Soviet diesel generators took a few minutes to power up, and during that gap, scientists wanted to find out if the residual power generated by a turbine that was still turning would be enough to power the pumps that kept cooling water running through the core of the reactor until the generators could take over.

However, design flaws in the reactor itself basically set the reactor up for an accident if the right things took place, and on that particular morning, those things happened. Although the accident was officially blamed on operator error, there were many things that went into creating the Chernobyl disaster, including the Soviet method of doing things.

How The Accident Went Down

Soviet officials were determined to have the turbine test run to check out whether or not the turbines could give a few moments of electrical power in the event of a power failure at the plant. The test should have been run when the reactor was new, but due to the Soviet penchant for rewarding those who got projects finished early, the test was postponed until a time when reactor conditions were no longer optimal in terms of fresh fuel and prepared operators.

The test was supposed to be performed by the afternoon crew, but the power authority asked them to postpone the test until the night, when power demands were less. So the afternoon crew, which had been prepared for the test, gave way to the night crew, which knew little about the test and were not prepared to conduct it.

The turbine test involved powering down the reactor to imitate an electrical blackout, and workers also shut off the alarm and emergency systems that would inevitably go off when the reactor powered down. Because the fuel was older and had built up a lot of decay products, the reactor became unstable when powered down.

Moreover, as the reactor powered down and the turbines failed to generate enough electricity, the flow of cooling water through the core slowed greatly. Since most of the safety systems had been shut off for the test, no extra core-cooling water was forthcoming from the emergency pumps.

Operators realized the reactor was becoming unstable and frantically tried to bring the reactor back up to full power, but when they did, the reactor surged, putting out more power than it was geared for. One operator saw what was happening and hit the SCRAM button to shut the reactor down by lowering all the neutron-absorbing control rods into the core.

The dashboard showed the control rods hadn't moved far before they stopped and loud knocking sounds were coming from the reactor hall. The operator quickly hit the release that would drop the control rods into the core by their own weight. But they didn't move.

This was because a major design flaw of the reactor suddenly made itself known. The tips of the control rods were made of graphite rather than boron, and when those tips entered the core, the power surged even higher because the graphite increased the reaction. The core, already red-hot, turned the small amount of cooling water flowing through instantly into steam and the pressure quickly built until a mighty steam explosion occurred.

Immediately after that, a second, far more powerful explosion occurred. Experts are unsure whether the second explosion, which blasted an estimated fifty tons of vaporized nuclear fuel high into the sky along with some 700 tons of graphite from the core that were scattered over the plant premises, was another steam explosion or a hydrogen explosion. Or it may have been a mixture of both, as the first explosion blew the heavy biological shield off the top of the reactor, exposing the core to inrushing air and oxygen.

The two deaths that happened immediately at Chernobyl were due to the explosions. One man, whose body was never recovered, was in a room next to the reactor. The other was also close to the reactor and was so badly burned and irradiated, he died in the hospital the day of the accident.

The presence of oxygen also ignited a fire in the graphite core remaining, which burned for several weeks. The fire added more radioactive isotopes to the environment over the period of time it burned.

It's Worse Than You Think

Meanwhile, back in the control room, the Deputy-Chief Engineer, who was in charge, decided the explosion had been a mere hydrogen explosion in the emergency water tank, and that the reactor was still intact. According to Andrew Leatherbarrow, writing in *Chernobyl 01:23:40*, "If he had looked out the window he would have seen that he was wrong." Moreover, the engineer told this belief to anyone who asked, and the authorities in Moscow believed it for almost a whole day.

In truth, anyone who looked outside at the destruction would have realized immediately that the reactor was destroyed, and that the chunks of smoking black stuff littering the plant grounds were pieces of the core.

Operators Disbelieve Their Own Instruments

After the explosion, one of the plant workers took a reading with his instrument. The instrument promptly went off-scale. Since the device was a very low-level meter, designed for use when all was well with the plant, the highest reading he could obtain was 1,000 microRoentgens. In normal times, this would have been considered a high reading.

Since the instrument was off-scale, the worker had to take a guess, and he guessed 5 Roentgens per hour, which was nowhere close to the real (much higher!) reading.

When the Chernobyl plant manager, Viktor Bryukhanov, arrived on the scene, the only working radiological meter the dosimetrists (workers charged with measuring radiation using radiological instruments) could find was another that measured only up to 1,000 microRoentgens per second. This amounted to 3.6 Roentgens per hour, and that was the reading the Deputy Chief Engineer in the control room and the plant manager chose to believe.

Believe it or not, the meters capable of reading the high levels of radiation from the accident were either buried under rubble or locked in a safe. The radiation sensors around the building itself had been destroyed by the explosion. They had no suitable meters on hand to detect the real levels radiation!

In reality, according to *Chernobyl 01:23:40*, by Andrew Leatherbarrow, readings in some parts of the plant reached 8,000,000 microRoentgens per second, or 30,000 Roentgens per hour. A lethal dose of radiation for humans is considered anything above 600 Roentgens per hour.

Then a staff member located a meter that read up to 200 Roentgens per hour, but it, too, promptly went off-scale.

Leatherbarrow writes, "Bryukhanov declared the device broken and refused to believe it." Both he and the chief engineer ignored other staff members who also took readings with instruments capable of higher readings, dismissing the meters as "worthless junk" and the workers as "fools."

The truth was they did not want to believe any reading above 3 to 5 Roentgens per hour. A high radiation reading would have meant what they did not want to even think about: either a major, career-ending accident or a destroyed reactor...or maybe even both.

Nobody Knows What's Going On

At Chernobyl, what was really going on was plainly visible to anyone who looked outside and saw the destroyed reactor building and the chunks of smoking graphite and nuclear fuel covering the ground. However, the two men in charge could not accept what had really happened.

In consequence, everything they did to try and "save" an already destroyed reactor made everything worse. Since a nuclear reactor must have plenty of cooling water flowing through, they got busy trying to feed water to the blown-out reactor. These actions exposed many of the heroic workers to high doses of radiation that resulted in some of the deaths associated with Chernobyl.

The workers in the control room were largely in shock after the explosions. No one could understand what had happened since they had done everything right, by the book, and everybody knew that nuclear power was so safe, you could build a reactor right in the center of Red Square and all would be well.

So just what had happened? Nobody knew for sure, and nobody really wanted to find out.

First Responders Are Ill-Equipped

Because the Soviet Union was on a very tight budget and there were constant shortages of almost everything, the nearby fire department had not prepared with storage closets full of respirators or other protective gear.

The air was loaded with radioactive particulates, but workers at the plant had nothing to help shield them from breathing in radioactive particles, nor did the firefighters.

Firefighters had no training for nuclear accidents, nor did they have the kind of protective gear they needed to fight the fires at Chernobyl. Many firefighters developed radiation sickness within an hour or two of firefighting and had to be taken to a hospital, where some later died. Most of them knew little to nothing about radiation and did not realize they were breathing in radioactive particles, while, at the same time, the open reactor was bombarding their bodies with extremely high levels of radiation, 30,000 Roentens per hour in some areas. They were told in the hospital that they'd been exposed to poisonous gas.

To the firefighters, a fire was a fire and they fought it as such. Some of the firefighters probably did realize they were fighting a dangerous radioactive fire, but not one of them shirked what they saw as their duty. Their heroic actions that night kept the fires that were started by the flaming graphite landing on the tar-covered roofs of the reactor buildings from spreading to the plant's other three reactors and probably causing meltdowns in them.

The flammable tar on the roof was a big no-no, but because the Soviet system rewarded those who finished projects on time, and because there were constant shortages of needed materials, Viktor Bryukhanov, who oversaw the construction of the plant, used the tar because it was available. He could not get the proper fireproof roofing material fast enough or in big enough quantities to bring the new reactor into production on time.

No preparations for accidents had been made at the plant, either. No safety drills had ever been performed. No respirators or other protective gear had been stockpiled on the plant grounds. How much of this was due to the belief that accidents couldn't happen at nuclear plants and how much was due to the usual Soviet shortages would probably be hard to estimate.

The Soviet hierarchy was so convinced of the safety of nuclear power that they actually believed (and convinced everyone else) that nothing could ever go wrong. This false confidence extended from the top all the way to the bottom tier of workers. Thus, when something did go wrong, blame had to be assigned and punishment meted out, but with no mention of the design flaw that underlay the accident. To do so would have called attention to all the other reactors in the Soviet Union built on the same design.

Rumors Abound

Perhaps the biggest rumor circulating around the Chernobyl power plant just after the accident was the one that swore up and down that the reactor was still intact. All it needed was a little water to help keep it cool. This was probably more wishful thinking than an actual rumor, and it did not last more than a day. In the end, the fact of the destroyed reactor had to be faced, along with the fact that the Soviet Union was not going to be able to keep this particular accident a secret from the rest of the world.

Not that they didn't try. When Sweden noted higher radiation readings at one of its nuclear reactors and determined that the readings came from the parking lot rather than the reactor, they checked the isotopes and traced wind directions before calling the Soviet Union to ask if there had been a nuclear plant accident. The Soviets denied it. However, other countries began detecting the fallout and denial soon became impossible, so they admitted an accident that was in the process of being mitigated, and that two deaths had occurred. At the time, this was the truth.

However, because of the Soviet habit of denial, nobody believed them. United Press International received an "insider" report from an informant in Kiev who said over 2,000 people had died in the accident, and that figure was printed all over the West. The *New York Post*'s headline read: "Mass Grave for 15,000 N-Victims."

This was "fake news" at its finest.

Who's In Charge?

Nobody doubted who was ultimately in charge at Chernobyl—the Politburo, the policy-making committee of the Communist Party.

In the Chernobyl accident, we saw no bumbling or stumbling as responders tried to decide who was in charge and whose orders to follow. Everybody knew who was really in charge—the leadership of the Communist Party—but in the Soviet system, this led to a whole new set of problems.

The Politburo was in charge, all right, but unfortunately, its members knew next to nothing about nuclear power plants or what the ramifications of the accident were. This caused them to deny to the world that an accident had occurred in Soviet territory, when nearly the whole world possessed evidence that one had in the form of radiation readings and isotope analyses that indicated a nuclear reactor accident rather than a nuclear bomb test. Consequently, when the Soviets did tell the truth about the number of deaths the accident had caused, nobody believed them.

In the Soviet system, mistakes, or errors in judgment, or design flaws in the Chernobyl reactor could not be admitted, as this might lead to people questioning the wisdom of the leadership. Therefore, they pinned the blame on the operators of the reactor and the manager of the power plant, when in reality, the design of the reactor almost guaranteed an accident sooner or later.

Aftermath

The Chernobyl reactor's explosion put out big plumes of highly radioactive material that was deposited in various areas all over Europe, some of which are still radioactive after thirty-plus years. Areas around the reactor, including the town of Pripyat where the plant workers all lived, remain evacuated to this day.

The Soviets held a big inquiry and show trial, at which they convicted those of the operators left alive, including the Deputy Chief Engineer and the plant manager Viktor Bryukhanov, of failing to follow procedures and of many other breaches of protocol.

As the Deputy Chief Engineer later noted, the big inquiry was conducted by the very people who had designed the faulty reactor, but if any problems had been identified with the reactor, the West would have demanded that every reactor of that type in the Soviet Union be closed down. That would have been a huge blow to the already shaky Soviet economy.

Twenty-eight more deaths occurred within days to months after the accident from acute radiation sickness in the firefighters and plant employees who went above and beyond ordinary heroism in trying to save the Chernobyl plant.

According to Wikipedia, in 1994, thirty-one deaths resulted directly from the accident among the reactor staff and firefighters. In 2008, the United Nations Scientific Committee on the Effects of Atomic Radiation placed the total number of Chernobyl deaths at 64 and said that number would probably rise.

We have no idea how many other people may have died of cancer or other conditions caused by overexposure to radiation. Their deaths are not included in the official Chernobyl death roster.

In the meantime, remediation began with an effort to stifle the fire still burning in the graphite core by using helicopters to dump tons of borate and sand directly into the reactor core.

The cleanup used the military as workers to hand-pick up the pieces of highly radioactive graphite and nuclear fuel that lay all over the grounds and roofs of the reactor buildings. Machinery simply would not work in those high radiation fields. In fact, much of the machinery used at Chernobyl became so radioactive, it was buried in trenches or abanoned in a field near the plant.

Over 250,000 men did the work robots could not do, some remaining only a few minutes in the highest fields of radiation until they reached their full "allowable" dose of radiation. Some received far more than allowable and died early deaths of radiation-induced conditions. The men doing this dangerous work were known as the *Liquidators*.

By December of 1986, the Soviets had erected a huge concrete building over the ruined reactor that was called the sarcophagus. By 1996, the sarcophagus had about reached its limit and a new structure was proposed and built.

The new structure is called the New Safe Confinement and was built around the sarcophagus. It is scheduled to be finished sometime in 2018, as high radiation levels have slowed construction.

Other Results of Chernobyl

One thing the accident—a relatively "small" accident involving one nuclear reactor—showed graphically was that no nation was ready for a nuclear war. Providing medical care for the victims of the accident and cleaning up after it proved that a single nuclear strike on any city anywhere would be almost impossible for a country to deal with.

The sheer costs of the accident in terms of money and manpower were enormous. Decontamination, caring for the victims, cleaning up the area and building the sarcophagus cost the Soviet economy billions of rubles.

The accident occurred soon after Mikhail Gorbachev came to power, and he used the incident to push his policy of glasnost, or transparency. However, the high costs of dealing with Chernobyl are considered one of the main factors in the ultimate collapse of the Soviet Union.

The accident is still a major expense in Belarus, which said in 2009 that it spends about a million dollars daily on the results of the Chernobyl explosion.

Many areas of Europe were contaminated with cesium-137 and other radioactive isotopes, but cesium-137 has a half-life of 30 years, which means that on the 30-year anniversary of the Chernobyl accident, a lot of cesium-137 still remained to cause trouble in various areas of Europe.

In Germany, hunters who kill wild hogs for food must have their kills checked for radiation. Some wild hogs have been found that were too radioactive to eat. The problem is caused by the fact that wild pigs love mushrooms, and mushrooms are known to concentrate cesium-137 from the environment.

In Sweden, a wild boar was shot and killed that had ten times the safe limit of radiation levels in its body. Wild boars have moved into a remote area of Sweden that suffered the worst fallout from Chernobyl, so the problem is expected to worsen.

In various European countries, blueberry jams made from wild forest blueberries were found to have fairly high levels of cesium-137. Most of the blueberries came from Bulgaria, which was known to have been contaminated by Chernobyl fallout, although the jams were bottled and marketed in other countries like France and Italy.

Another interesting little tidbit, found in *Safe as Houses? Ill Health and Electro-Stress in the Home*, by Cowan and

Girdlestone, is the use of Chernobyl fallout to cover up other possible causes of cancer.

The authors say that in 1995, "microwave radiations from the missile firing range and tracking installation on the Hebridean island of Benbecula were being considered as the cause of a ten-fold increase in cancers, rather than the politically more acceptable scapegoat of caesium fallout from the 1986 Chernobyl nuclear disaster."

Apparently, any entity performing activities that may have caused cancer has so far been able to blame it on Chernobyl's fallout. It looks as though that era may be just about over.

What Really Caused The Accident?

It took a few years before the main cause of the Chernobyl accident came out. The primary data about the disaster, in the form of instrument and sensor data, were not generally available in full. Ultimately, an International Atomic Energy Association report in 1993 placed the cause of the accident, not on the operators as it had for several years after the accident, but on the reactor's design.

But underlying all of this was the pervasive lack of a safety culture in almost every aspect of Soviet industry. A 1991 report to the USSR State Committee for the Supervision of Safety in Industry and Nuclear Power pointed out that there was no top-level responsibility or oversight of the nuclear industry on a consistent basis in the Soviet Union. This was how a faulty reactor design like the one in use at Chernobyl managed to get approved and built.

And of course the reason for all these deficiencies likely lay in the idea that nuclear power was so safe, the plant could almost run itself.

4: The Accident At Fukushima

March 11, 2011

The Fukushima Daiichi Nuclear Power Plant, located on the eastern coast of Japan, boasted a grand total of six boiling water nuclear reactors. On March 11, 2011, two of the reactors were shut down for refueling, four were producing electricity, and all six reactors had spent fuel pools that required constant water circulation to cool the fuel.

The accident began with a major earthquake just offshore, which generated an extra-high tsunami. The tsunami flowed into the Fukishima Daiichi plant, where it flooded the emergency generators that would have keep water circulating to cool the reactor cores. After the earthquake hit, the four reactors shut down automatically, but the nuclear fuel still had to be kept cool. With no cooling water circulating, all four active reactors suffered meltdowns.

The Fukushima accident is a prime example of an accident that "couldn't happen." It was a "beyond-design-basis accident," meaning the plant had been designed to withstand smaller earthquakes and tsunamis than the one which occurred. An accident this severe is thought to be so unlikely by engineers that it didn't need to be considered when designing the nuclear plant.

Another factor that made the accident so severe that it rates a 7 on the International Nuclear Event Scale is that it wasn't just one reactor that melted down. Four reactors melted down, with some spent-fuel-pool trouble thrown in for good measure.

Although nobody died of immediate radiation exposure in the Fukushima accident, authorities attribute 1,600 deaths, mostly in elderly people who had been living in nursing homes, to the chaotic evacuations of towns and villages near the damaged plant. Some of these old people were evacuated more than once, which was a major stress on them.

On September 5, 2018, Reuters reported that "Japan has acknowledged for the first time that a worker at the Fukushima nuclear power plant ... died from radiation exposure." The man was diagnosed with lung cancer in 2016, after working at nuclear power plants all his life. He worked at the Fukushima Daiichi plant "at least twice after the March 2011 meltdowns at the station." He was in his 50s.

How The Accident Went Down

On March 11, 2011, at 2:46 p.m., an earthquake occurred about 80 miles offshore of the northeastern Japanese coast of Honshu. It had a magnitude of 9, a very powerful earthquake that generated a tsunami of about 43 feet high within 50 minutes of the earthquake. The seawall of the Fukushima Daiichi nuclear plant was only 33 feet high, and water flowed over it and flooded the lower lying rooms that housed the emergency diesel generators.

The reactors shut down immediately after the earthquake, as they had been designed to do, but they still needed the cooling water flow. When the electricity failed, the reactors relied on the cooling water pumped by those generators, but when the generators were flooded by the tsunami, they quit working. The inflowing seawater also damaged the electrical distribution system of the plan, and all electricity was out.

Without a cooling water flow, a reactor's core begins to overheat within 30 minutes of losing cooling, and deformation and melting of the fuel rods begins soon afterward.

It's Worse Than You Think

Workers at the plant knew this was a bad situation and immediately brought in some battery-powered generators. These generators were intended to cover any gaps in the electrical power and the diesel-generated power. The batteries weren't very long-lived, just eight hours, but then, they didn't have to be. Power outages never lasted very long, after all.

However, what happened on March 11, 2011, was a scenario that no one had ever dreamed could happen. It set the stage for a nuclear accident in several reactors at the same time, something that no one had thought possible, including the design engineers.

First, power could not be quickly restored this time because of the extensive earthquake and tsunami damage to the infrastructure. Second, fresh batteries, workers and supplies could not be brought in because of debris and damage to the roads. The current workers were on their own, with what they had on hand.

The workers tried valiantly to overcome the shortages by raiding any cars and busses, even their own cars, of their batteries and attempted to wire them together. They ran hoses from the Pacific Ocean and attempted to pump seawater into the cores. They did everything they could think of to get water to the cores before a meltdown could occur, but the extensive damage to the plant and its electrical system kept them from achieving success.

Meanwhile, Back In Tokyo ...

Nuclear energy in Japan was overseen by various agencies to a greater or lesser degree. The government thought it had everything covered, until the accident happened and the oversight system failed to work. This was not a really good time to discover that the functions of these agencies often overlapped or conflicted, which meant that, once again, *nobody knew who was in charge.*

Tokyo Electric Power Company (TEPCO), the plant owner, declared an emergency, a notification meaning that an accident either had occurred or was predicted. Its next step was to notify the Ministry of Economy, Trade and Industry, the governor of Fukushima Prefecture and the mayors of two towns near where the plant is located, by fax as required by law. With no phone service and no electrical power in those areas affected by the earthquake and tsunami, that could not be done.

When a nuclear plant accident occurred in Japan, regulators from the government's Nuclear and Industrial Safety Agency were supposed to come to a command post off the nuclear plant site to help with the emergency response, but the command site was located three miles from the reactors. With no power, no phones and no food and water, the regulators who arrived could do little. Worse, air filters had never been installed at the command post in case of a radiation cloud, even though the agency had been cited twice already for that failure by inspectors.

No Communications

Even the workers at the plant could not communicate with each other because their emergency radios had short-lived batteries that could not be recharged without electricity. A worker with a message had to hike quite a distance around the plant in order to deliver his message.

The communication problem extended to Tokyo. Even though two groups of authorities had gathered on different floors of the same building to deal with the accident, they had no communications with each other. Nor did any of the authorities, including TEPCO senior managers, know what was happening at the plant.

And Back At The Plant ...

Things were not much better at the Fukushima Daiichi plant. With the power failure and the destruction of much of the electrical circuitry, the workers were receiving no signals from the reactors about what was going on inside them. All those little dials and instruments in the control rooms failed to work without electricity.

Nobody knew what the water levels were inside the reactor cores. The control rooms were totally disabled by the loss of power, and the usual constant flow of information about a reactor's status had now become a guessing game. The one thing workers knew for sure was that if they did not manage to get a flow of cooling water started, all the working reactors would suffer meltdowns.

They did what they could with their scavenged car batteries and sent the plant's three fire trucks to the reactors in hopes of using them to get water into the reactors. But everything that could go wrong went wrong, and as pressure inside the containment vessels rose, operators knew they had to vent some of the pressure in order to introduce water.

Venting from a damaged reactor can release radioactive contaminates into the atmosphere, and it is basically an admission that things are desperate. Any venting had to be approved by various entities, including TEPCO's chief executives, who were out of the country, and the Japanese government. Needless to say, a decision could not be reached quickly.

First Responders Are Ill-Equipped

Back at the plant, workers kept discovering that their emergency manuals had omitted a few major items, including how to operate the venting valves by hand, as there was no electricity. Then there was the problem of *finding* the valves that could be opened manually. It appears the emergency manuals failed to include this bit of information, also.

All this chaos played out against the major tragedy that had hit Japan in the form of the tsunami that had washed away entire villages and had killed almost 16,000 people, with around 2,500 listed as missing. In these first hours after the tsunami, people still awaited rescue from atop buildings where they had fled the waters, and the cleanup facing the authorities was horrendous. So it is hardly surprising that the Japanese government was overwhelmed and did not immediately recognize the seriousness of the situation at Fukushima Daiichi.

The government was late in declaring the nuclear emergency to the nation and the world even as radiation levels inside the Unit 1 reactor rose so high workers could no longer enter the building. Nobody knew what the water levels were in Unit 2, but they feared the worst.

In the evening at long last, mobile generating trucks were dispatched to the plant, although it took them a long time to make their way over roads that were crowded with fleeing survivors and tsunami debris. But when the trucks arrived late that night, their cables were too short and their plugs were not compatible with those at the plant.

Nobody Knows What's Going On

At the plant, with no electricity and no working panels of instruments in the control rooms, none of the workers knew the states of the reactors, although they knew Unit 1 was in trouble, and they figured Unit 2 might be in trouble. They had none of the usual information from the control panels available to them, and their emergency guides seemed to be missing a lot of crucial information.

In Tokyo, where TEPCO's chief executives were still unavailable, both TEPCO and the government agencies knew even less than the workers at the plant. But each agency knew something different, which served to keep things interesting.

What Instruments?

In the case of Fukushima, the workers would most likely have believed their instruments, if only they had been working. With no electricity, all the control panels were out of action. The instruments that were working, namely the radiation meters, told the workers things were steadily growing worse, and they had enough sense to believe the readings in light of the current situation.

Back in Tokyo, the TEPCO managing director, and the heads of two of the government agencies charged with nuclear plant oversight held a press conference to announce that a venting of gas buildup would take place at Fukushima Daiichi.

Before they went on camera, the three men discovered they each had different information about the state of the reactors. They agreed on a strategy of not identifying the reactor that would be vented, but press questions soon had TEPCO's managing director thoroughly confused.

The Japanese people noticed the confusion.

While the press conference played out in Tokyo, back at the plant, the Unit 1 reactor got tired of waiting and vented itself, quite literally by starting to come apart at its seams from the high pressure inside.

The workers now had lower pressure in the reactor vessel so that they could inject water into the core, but then they discovered nobody knew how to operate the fire engines. They persuaded a frightened contractor to help. After all, radiation levels at the plant were rising and the earth beneath their feet continued to roll with aftershocks, which meant another tsunami could be likely at any moment.

Alas, by the time they managed to get some water flowing into the Unit 1 reactor, it was too late. The core by then had most likely already melted down and vacated the reactor vessel.

And so it went for the first twenty-four hours after the tsunami had first hit.

Everything the workers tried to do to mitigate things seemed doomed to failure because of unforeseen problems that nobody had ever addressed in the emergency plans, because nobody thought an earthquake that big could cause a tsunami that high that could cause such damage that four reactors would melt down at the same time.

About 24 hours after the tsunami had flooded the plant, the Unit 1 reactor building literally blew its stack. A powerful hydrogen explosion blew off the roof and scattered debris over the area.

At this point everyone in Japan and around the world realized that Fukushima Daiichi was a major catastrophe in the making.

It Kept On Going ... And Going!

Fukushima dominated the news in America and probably around the world for several weeks after the tsunami and earthquake. The explosions clearly meant that fuel damage and meltdowns had occurred, and not just in one reactor.

The fact that Unit 3 suffered a hydrogen explosion on March 14, followed by an explosion at Unit 2 and a hydrogen explosion at the Unit 4 building that collapsed the top two stories made the situation even more dire. The Unit 4 explosion meant that the spent fuel pools at the other reactors were also in danger. Things at Fukushima Daiichi seemed to deteriorate by the hour.

The Japanese government had ordered many small villages, already devastated by the earthquake and tsunami, to evacuate. When things deteriorated rather than improved, they widened the evacuation. Many bewildered, traumatized people who had evacuated once already found themselves being evacuated again.

As radiation levels at the plant began to rise, the government came up with a clever solution to the fact that the plant workers had almost reached their allowable limits for radiation exposure—they raised the allowable levels.

Meanwhile, In America ...

American authorities experienced the same trouble Tokyo was having in trying to find out what the heck was going on at the plant. That is hardly surprising, because nobody else knew either, including the people at the plant.

Tracking any radiation plumes released from the plant was rendered difficult because the plant had no electricity. That meant no information from the plant about any release could be transmitted to Japan's system of tracking radiation releases.

On top of that, the Nuclear Regulatory Commission (NRC) wanted to recommend a 50-mile evacuation zone for Americans living in Japan near the reactors, based on what they knew at the moment. They were held back because (1) the Japanese government was only recommending a 12-mile evacuation zone, which would cause diplomatic problems and (2) this would bring America's own 10-mile-zone recommendations around American nuclear reactors, including the 23 reactors built on the same design as those at Fukushima, into question.

For a while there, it looked as though 4 nuclear reactors were out of control and melting down, and 6 spent fuel pools were in danger of going dry. How bad could it get?

Nobody knew.

Rumors Abound

Because of the lack of information on what was really happening at the Fukushima Daiichi plant, any guess, no matter how wild, could not be ruled out. Nothing like this had ever happened before. It was supposed to be impossible.

In America, people went wild trying to stockpile potassium iodide. Radiation was coming, on the ocean currents and the air currents. Health food stores in Texas that happened to have a supply shipped the pills to people in New Jersey, as people all over America scrambled to find the pills anywhere they could, and online supplies had long ago sold out.

People on the West Coast monitored the air and the water, and some reported normal levels of background radiation as proof that Fukushima contamination had already reached American shores. Scientists calculated the possibilities as to how much radioiodine could soon turn up in American milk supplies. Parents worried over their children, and pet owners worried about giving potassium iodide to their pets.

Passengers flying into America from Japan were screened for radiation, as if the whole of Japan lay beneath a nuclear cloud.

In China, word was that a big radiation cloud was moving in, so people rushed to grocery stores and bought up all the iodized salt, which was said to protect against radiation. Others bought up all the sea salt, believing that all sea salt from now on was likely to be contaminated with radioactive isotopes.

And in countries located near Japan, people worried about the ocean currents, the air currents and the food. Many of those countries imported rice and other foods from Japan. All of a sudden, foods of Japanese origin stopped selling, even though they had been on the shelves before the disaster, another indication of how little people understand about radiation.

Distrust of Government

In Japan, things were no better. Citizens had quickly realized that nobody in charge knew what was happening because of the contradictory information that kept coming out, and they totally lost faith in the government. So far as the Japanese people were concerned, the government was hand-in-hand with the nuclear industry and intended to soft-peddle any serious concerns.

Worried and seriously mad, Japanese citizens searched for better sources of information.

The people bought up every radiological meter in Japan within a couple of weeks of the earthquake and started monitoring areas around them on their own. Government readings were ignored in favor of citizen-run websites that posted readings taken at schools, in neighborhoods and other places people tended to go.

Operation Safecast, a group of Japanese citizens with radiological meters, began posting readings for certain areas regularly. Although the government tried to undermine the operation, the people trusted Safecast readings because those readings told them what they wanted to know about the areas they frequented.

When the Fukushima plant experienced explosions and radiation releases, the people soon realized government information told them nothing useful about any radiation plumes, assuming it wasn't contradicted a few hours later by the next newscast. Instead, they sat over maps in their own kitchens plotting wind direction like professional meteorologists.

When people shopped for food, they checked it out with their radiological meters before they placed it in their carts. Some stores got with the program and actually provided radiation readings for various foods, posted on signs near the food.

Evacuations—An Exercise In Confusion

When things at the Fukushima Daiichi plant began to deteriorate and major radiation releases became likely, the Japanese authorities immediately announced an evacuation of the area within 12 miles of the plant. The people being evacuated were already shell-shocked from the twin major disasters of the earthquake and the tsunami, and now they had to move further, and they did not know why. No information was given them except to pick up their belongings and start moving.

Immediately after this, things worsened further at the plant, and the people who had moved 12 miles from the plant were told to move again, this time 18 miles from the plant.

The stress from these repeated movements was bad enough, but what made everything worse was that many of the people were moved into areas that may have been far enough away, but they were in the direct path of the radiation cloud. Many people were literally moved from the frying pan into the fire—they'd have been better off staying where they were.

In fact, the deaths that came from the Fukushima Daiichi disaster were not a result of radiation sickness or explosions at the plant. They occurred in elderly people residing in nursing homes who became highly stressed from being moved several times, and in one famous case, being left behind in a facility for a couple of days with no care and no electricity.

Other people were told to shelter in place, to stay inside their homes and do not come outside. So they did...until they realized no trucks would deliver food to their area.

Can A Government Really Protect Its People?

All-in-all, the government response to the crisis was described as "chaotic and leaderless." So we can't be surprised that the Japanese people reacted accordingly, taking matters into their own hands.

Information was deliberately withheld from the public in order to "avoid panic." Naturally, the people realized this very quickly and lost faith in the government. The government, as if to confirm its lack of trustworthiness, did not even admit the three meltdowns that occurred in mid-March until June. In fact, the word *meltdown* was stricken from the vocabulary of TEPCO officials appearing on television.

Who's In Charge?

Meanwhile, back in America, the Nuclear Regulatory Commission (NRC) was having its own troubles. It had problems with protecting its turf from the Department of Energy, and it had even more trouble trying to figure out who the players were in Japan among all the ministries, TEPCO officials and government officials.

It soon became very clear to the NRC that in Japan, *nobody was in charge*. It took the NRC representatives who had been sent to Japan several weeks to figure out which, among all the government agencies and company men, were most likely to know what was going on at the plant and could be most helpful to the NRC men.

As the Japanese people had already noticed, the whole response to the disaster seemed "leaderless" and disorganized.

Furthermore, it became more and more evident that the NRC had little ability to give any useful advice to people in America who were worried about the incident in Japan, especially in a case where it couldn't even find out what was happening. Moreover, it had never planned for an incident that (1) involved multiple reactors all melting down at once, and (2) involved a situation that got worse by the minute because everything that could go wrong, did.

As we all know, the problems at Fukushima Daiichi are ongoing, and it will be another 30, 40 or more years before the plants can be totally shut down and cleaned up. Leakage of contaminated water into the Pacific Ocean continues to cause trouble, and nobody has any idea where the melted fuel in the reactors got off to. The reactors remain far too radioactive for even a robot to survive.

What Was The Real Problem at Fukushima?

The real problem at Fukishima wasn't the huge earthquake, or the enormous tsunami that hit. In the end, the real problems at Fukushima Daiichi boiled down to two things:

(1) The fact the government officials responsible for regulating the nuclear power plants deferred to the needs of the nuclear power industry and worked to advance the agenda of the nuclear power industry. In short, the regulators were hand-in-hand with the regulated. It was later shown that the owners of the plant knew a bad enough earthquake or tsunami could cause big trouble at the plant, but did not want spend the money to make the necessary upgrades.

(2) The overriding belief that an accident this bad just could not possibly happen. This belief probably underlay almost every decision and regulation made concerning nuclear power before the Fukushima Daiichi accident.

In fact, we can safely conclude that all three accidents we have looked at can be partly blamed on (2).

At Three Mile Island, at Chernobyl and at Fukushima Daiichi, no one really believed an accident could actually happen, and certainly not a bad accident.

After all, everyone knew that the plants had instituted redundant safety measures. With all those safety protocols in place, how could any accident possibly happen?

Part II: Lessons Learned

5: It's Worse Than You Think

We saw, in all three accidents, that when instrument panels first displayed unusual readings, workers did not immediately grasp the implications. In fact, as the readings rapidly grew worse, workers had a tendency to assume it was the instruments that had gone haywire, not the reactor.

Probably Normalcy Bias has a lot to do with this. Workers who have worked for months or years at their jobs without ever experiencing an alarm or an "incident" come to believe that all those safety precautions and redundant systems designed to kick in when something goes wrong are doing their job. If anything happens, it will quickly be resolved.

In the end, today will be like yesterday, and tomorrow will be like today. If we had no accident today or yesterday, then we won't have an accident tomorrow. This is a dangerous assumption, especially without solid proof.

Consequently, when an accident does occur, workers experience a great reluctance to admit that it is an accident, and they are even more reluctant to entertain the thought that it might be a bad accident.

Even with things around them visibly deteriorating, with alarms shrieking and explosions rocking the control room, workers usually believe the cause is nothing serious. Often, they come up with explanations for the alarms that have nothing to do with the real cause.

Always assume the situation is worse than you think!

Three Mile Island

At Three Mile Island, workers knew something unusual had occurred, but because they could not see the instrument that would have helped them discover the problem, they kept on assuming that something non-serious was happening.

Even when the radiation alarms began shrilling because radioactive water was being pumped into the auxiliary building, nobody realized the reactor was rapidly losing its cooling water. In fact, they assumed the opposite!

Moreover, despite all the signs, such as the hydrogen bubble and the high radiation levels, nobody knew (or would admit) that a meltdown had occurred. *The meltdown was not admitted until three years later.*

Chernobyl

When the Chernobyl No. 4 reactor became unstable, one worker immediately initiated a "scram," where all the control rods are inserted into the core. Rather than shutting down the reactor, the action instead increased the power exponentially, flashing any water remaining in the core to steam. Steam pressure built up until the reactor finally exploded.

Even after the explosions, the workers did not immediately realize how bad the situation was, perhaps because they didn't really want to realize it. The Deputy Chief Engineer decided the explosion had not been in the reactor—it had been a hydrogen explosion in one of the auxiliary water tanks. The reactor was okay. It was still intact and just needed more cooling water.

A pair of young workers died because they were sent to check on the reactor. When they went to look, and actually looked down into the partially open reactor, they received such a strong dose of radiation in that minute or two that their faces developed an instant "nuclear tan." They were among the 23 deaths that occurred within a few weeks of the accident.

But when they reported back that the reactor had been destroyed, the Deputy Chief Engineer did not believe them. He continued to insist that the core was intact, and all it needed was water.

Not even the plant manager, Viktor Bryukhanov, wanted to hear that the reactor was destroyed. He accepted the Deputy Chief Engineer's assessment as fact and passed it on to Moscow, even though he had seen the damage for himself upon arriving at the plant.

Both these men also accepted the fiction of low radiation levels at the plant, even though they, above all other people, knew that those low levels were the highest the available radiological instruments were capable of reading.

The personnel running the plant weren't the only ones hoping the situation wasn't at all bad. The Politburo in Moscow, the top men of the Communist Party in the Soviet Union, had a really tough time coming to grips with the fact that they were not going to be able to hide this accident the way they had others.

They understood little about nuclear science, and the nuclear scientists who did understand the worldwide ramifications of the accident had to do a lot of talking to convince the politicians that this accident could not be hidden from the rest of the world.

The situation at Chernobyl was, indeed, far worse than the politicians of the Politburo thought and the plant manager hoped.

Fukushima

Fukushima was probably the only accident where the workers likely had a very good idea of what was going to happen if they could not establish a cooling water flow to the working reactors. But they, too, probably had a hard time coming to grips with the fact that in spite of all their heroic efforts, they could not outrace the clock and get cooling water to the overheating cores of the three reactors in time to prevent meltdowns.

The interesting thing about the Fukushima accident was the steady downhill decline of the reactors, no matter what action the workers took. It seemed that whatever could go wrong did go wrong. Every action the workers took seemed to suffer an unexpected glitch, so that rather than improve the situation, things continued to deteriorate.

Not even the emergency manuals that supposedly contained detailed instructions for dealing with emergencies could help. It appears the writers had never sat down and thought through what would happen in a really bad accident, so the workers were more or less left to fend for themselves.

Very likely, at first the workers had felt they would be able to deal with things, since they had those detailed emergency manuals that would surely cover any contingency.

At Fukushima, everything about the situation was worse than anybody thought; worse even, than anybody had ever dreamed.

6: Nobody Knows Anything

When an accident occurs and everybody realizes something untoward is happening, the chief emotion seems to be an awful confusion. Nobody knows what's happening, or what is going to happen, or even what has happened. The event is outside their training and experience.

This state of being can go on for quite a while. Often, it seems to end only when someone finally gets a grip on what's going on and spells it out for everyone else.

How much of this confusion is caused by the natural tendency of people to believe that the situation is not as bad as it might look is hard to say. But those involved seem to remain in a state of confusion until the situation becomes clear to at least one person, and suddenly everybody knows what's going on. Often, this occurs too late.

Unfortunately, the news media usually arrive while the people involved are still in the state of confusion. Many of the stories reported may be contradicted by other stories, and the reporters soon develop a cynical mistrust of the available sources of information.

Three Mile Island

At Three Mile Island, for the first few hours everybody involved was in a state of confusion. Something was wrong, and whatever it was caused the radiation alarms to shriek and the warning lights on the control panel to blink on, but all their training had not prepared them for the situation that developed early on the morning of March 28. The instruments gave confusing information that left workers baffled.

It fit no scenario they had been trained to handle.

The primary cause of confusion lay in the fact that they had no instrument that told them the amount of water in the core. They inferred that information from other gauges, but in this particular instance, the information from those gauges conflicted. The operators thought anything and everything, other than that the core of the reactor was actually suffering a loss of cooling water.

Unfortunately, this state of affairs went on for quite a while, even after the open valve that lost all the cooling water was discovered and closed. It was compounded by the arrival on the scene of Nuclear Regulatory Commission representtatives, who also knew nothing.

The NRC representatives had a hard time finding out what was going on, and since the press relied on them to know something, *it soon became very obvious to the reporters that nobody knew anything.* One representative contradicted another, who contradicted officials from the state, who contradicted executives of the company that owned Three Mile Island.

The press could only draw one conclusion: Nobody knew what was going on, including the people who were supposed to know. *That meant the situation was really dire.*

Chernobyl

The powerful explosions at Chernobyl were such a shock to the workers, it is hardly surprising that nobody knew what had happened at first. The engineer who pressed the "scram" button was especially baffled. He had done everything right, by the book, and he could not understand why the explosion had occurred.

In fact, many of the plant workers thought the plant had been attacked, that the Soviet Union was now at war. Considering the state of the plant, that probably seemed a logical conclusion at the time. Especially since everybody knew that nuclear power was so safe, an accident could not happen.

The plant manager and the deputy chief engineer clung like barnacles to the mistaken belief that the reactor core was intact and the explosion had occurred in an auxiliary water tank. This was what they told the higher-ups in Moscow when they called to report the explosion.

In the Soviet Union, nobody had to worry about explaining things to the press, but they did have to worry about what they would tell the top officials of the Communist Party.

The Communist Party officials, in turn, worried about how they were going to hide this event from the world.

The event was so unprecedented, no one knew what to do about the reactor, which was still burning and spewing radioactive particulates into the air.

Worse, if they evacuated areas around the plant, that was tantamount to admitting the accident was a bad one, and that the situation was really dire.

Fukushima

Fukushima was one of those events that just kept on cooking, no matter what was done or not done. At the plant, the workers knew what was happening, but there wasn't much they could do. The disaster had rendered all the backup systems inoperable, and they didn't know what to do about it because all their training had not prepared them for such an unheard-of disaster.

Everything the workers tried to do seemed doomed to failure. But you had to hand it to them, because they hung in there and persevered, even when the hoses they'd spent hours laying were damaged by debris when Unit 1 blew its top and they had to start all over again.

The thing causing confusion for the workers was the lack of training for such a worst-case scenario and the wholesale failure of the severe accident manual to give clear instructions for the situation that existed.

At Fukushima, the worst confusion appeared at the government and TEPCO headquarters in Tokyo. Because of the unusually severe natural disaster, communications were down. Finding out what was going on at the plant was difficult, considering conditions there.

Then there were the overlapping jurisdictions of the various government agencies that were supposed to oversee nuclear power plants. When push came to shove, the whole system proved inoperable, creating much confusion in the meantime.

Then when representatives of the American NRC came to Japan to help out and offer advice, they had an awful time finding out what was going on. Nobody seemed to know, and if someone did know something, another person contradicted him. It took them weeks to find out which of the many agencies and TEPCO officials involved were most likely to have the latest and most accurate information.

At Fukushima, nobody doubted that the situation was dire. They just had no idea how dire it was, or how dire it might become.

7: Operators Disbelieve Their Instruments

When a real crisis occurs, operators seem to really, really want to believe that it is the instrument that is wrong, not the item the instrument is measuring.

This is understandable. In the confusion surrounding an event of some kind, even if the nature of the event is still unknown, nobody wants to believe that radiation levels are rising. That would mean something really serious has occurred.

Three Mile Island

The operators at Three Mile Island knew something was wrong because the instruments in the control room were blinking and beeping alarms, but they could not figure out what the problem was.

One of the computers printed out temperature readings from the core of the reactor, which showed the temperature steadily rising in the core. Or perhaps the thermocouples that measured the core temperatures were acting up.

When a worker bypassed the computer using a voltmeter hooked directly to the thermocouple wiring, he obtained a reading of 2,400 degrees. Every reading he took showed the core temperature at more than 2,000 degrees.

Operators immediately assumed the thermocouples had gone kaflooey.

Everybody assumed the information from the thermocouples was inaccurate, and nobody gave any credence to the idea that a core meltdown was occurring.

Chernobyl

At Chernobyl, where they should have known something major had occurred, given the magnitude of the explosions, everybody still wanted to believe the reactor was intact. But high radiation readings would have meant something bad had happened with the reactor.

Added to this was the fact that the explosions had destroyed the powerful radiation sensors deployed around the reactor building, and the only two radiological meters capable of measuring the really high levels of radiation actually present were locked in a safe or buried in the rubble.

Consequently, when a dosimetrist took a reading with one of the very low-level instruments they usually used every day (it was all that was available) and it went off scale, the real reading could have been almost anything. So the dosimetrist was left to take his best guess. It isn't surprising that his best guess was also very low, far below the actual levels of radiation present.

When a couple of workers turned up some high-level radiation meters, the plant manager, who was reporting to Moscow, did not want to hear them. His entire career was on the line and he knew it. Probably, he wanted to believe in the water-tank-explosion scenario as long as he could, no matter what his own eyes had told him when he first arrived at the plant.

Instruments are only good if the information they give is accurate and acted upon.

Fukushima

The problem at Fukushima Daiichi wasn't that the operators disbelieved the instruments—it was that the instruments were out of commission and told the operators nothing.

They knew that without electricity, and without the powerful generators that took over when the electricity failed, the reactor cores would begin to melt down if a flow of water was not reestablished within a few hours. Their problems with achieving that came about because the electrical system had been damaged by the tsunami, among other things.

Most of the problems these workers had lay with the government and TEPCO management, both of which attempted to manage events from a distance. The venting could not be done until the government okayed it, and by the time they were finally told to vent, conditions at Unit 1 reactor had deteriorated so much, workers could no longer go inside. The reactor wound up venting itself.

According to *Fukushima: The Story of a Nuclear Disaster*, the plant manager and workers were "besieged with orders and counterorders from TEPCO headquarters, which was also getting instructions from the prime minister's office." These often conflicting orders delayed actions that needed to be taken at once, and would have created chaos among the workers had not the plant manager ignored many of them.

The Japanese people also showed excellent sense. When they failed to get adequate information from the government, they bought meters and began monitoring their surroundings on their own.

In the end, instruments must be kept in good condition, their calibration must be maintained and the operator must know this so he can believe the reading he obtains.

Look for this problem of believing readings given by instruments to repeat in the near future when a nuclear incident goes down in America.

8: First Responders Are Ill-Equipped

In these disasters, first responders were often ill-equipped to handle the problems presented for various reasons, depending upon the incident.

But the major problem seemed to be the lack of a safety culture that would have led to greater preparedness. This appears to go back to the idea that nuclear power is so safe, that reactor designs have so many safety features and redundancies, that accidents are next to impossible.

Of course, that is the operative phrase—next to impossible. In each of these cases, the impossible happened and first responders had to make the best of their equipment and their training.

Also, first responders may not truly understand radiation and will thus take chances or go into areas they should avoid. Still others, particularly in America, may refuse to go into areas where radiation levels may be above normal.

Three Mile Island

The Three Mile Island accident fortunately never required the actions of first responders, although many were prepped and ready to go. We feel sure that if their services had been required, we would have soon learned of any inadequacies in their training.

In this case, the Pennsylvania Emergency Management Agency was prepared to respond, but as it turned out, they were not needed. Three Mile Island just missed being something really awful, and in a heavily populated area of the country.

Chernobyl

Chernobyl required firefighters from several nearby towns to come out and extinguish the fires burning on the roofs of the reactor building and nearby buildings. To say they were not prepared for what they had to deal with is understating the case.

The Chernobyl reactor had no containment vessel, which would have helped to prevent radiation from escaping into the environment. This simple precaution helped mitigate the Three Mile Island and Fukushima accidents.

The explosions at Chernobyl released huge amounts of radioactive contaminants into the atmosphere. These radioactive *plumes* were carried all over Europe by the prevailing winds and still contaminate some areas today. This toxic atmosphere is what the firefighters at Chernobyl faced, in addition to the dangers from the fire itself, and from their nearness to the highly radioactive reactor.

First, they had no respirators. Probably, they did not even realize they needed them. Immediately after the explosions, the air was loaded with radioactive particulates, many of them alpha-particle emitters and beta-particle emitters. These particles do most of their damage if they manage to get inside the body, and it was said that in the case of more than one firefighter who died within a few weeks of the accident, their lungs had literally been burned black by the alpha and beta radiation they had inhaled.

Second, they had no protective clothing. To these firefighters, a fire was a fire and they tackled it as they would any fire at any industrial plant. In the case of Chernobyl, it is difficult to say whether or not protective clothing would have even helped, especially for those firefighters who got on roofs anywhere near the open reactor. The radiation was so strong there, the only protective clothing that might have helped would have employed two-inch-thick lead sheets.

Third, they had absolutely zero training in fighting radioactive fires, and some ventured into the reactor hall to shoot water into the burning reactor. Here, they received fatal doses of radiation in less than one minute.

A major problem at Chernobyl was the total lack of a safety culture. To have proper equipment available, not to mention proper training for fighting radioactive fires, would have cost a lot of money. Worse, it might have suggested to the Soviet people that nuclear power wasn't quite as safe as the leaders made it out to be.

As Andrew Leatherbarrow points out in his book on Chernobyl:

> Incredibly, it transpired afterwards that no proper, full fire drill had ever been conducted at the plant. Even the procedure for fighting fire at Chernobyl was almost identical to any other industrial fire, with no regard for the possibility of radiation exposure—so presumptuous were senior figures that nothing could ever go wrong.

Even the doctors who responded to the plant to help with any injuries had no protection. The first doctor to arrive at the plant knew he needed a respirator, but there were none available. Nor were there any supplies for the types of sickness he was seeing. He remained on the job, however, even after he became ill himself because, "When people see someone in a white coat nearby, it calms them."

No doubt some of the firefighters understood the danger, and the doctors definitely did. But at Chernobyl, nobody turned tail and ran. Everyone did what they saw as their duty, come what may.

Fukushima

The accident at Fukushima occurred in a technologically advanced country that had the latest equipment available. The reactors were General Electric Boiling Water reactors, and the workers were well-trained. However, TEPCO was well-known for covering up safety problems and invested a lot of time and money in maintaining a public image that allowed it a lot of leeway in its operations.

The natural disaster that befell the area, however, exposed a huge lack of preparedness and planning for worst-case scenarios.

In this case, documents produced some time after the accident showed that TEPCO had indeed known that a large magnitude earthquake could cause a situation like the one it faced on March 11, 2011, but it chose not to implement any changes because of the cost involved.

A higher sea wall might have kept the tsunami from flooding the plant, and this was one of the major failings listed. Experts had warned that a really big earthquake could produce a much higher tsunami than the wall was built for.

In addition, the way the plant had been built, with the emergency generators located on a lower floor which was flooded, and the spent fuel pools on higher floors, which were hard to access when the pools required the addition of more water, made the problems that developed from the lack of electricity far worse than they needed to be.

As things went, the workers did the best they could with what they had, and even though they had some good equipment available, it proved inadequate in the emergency that presented. The batteries provided to run the generators in a worst case scenario did not last nearly long enough for the real worst case that occurred, nor did the workers' two-way radios, because nobody had ever planned for a situation where power was not restored within an hour or two.

It became pretty obvious that the writers of the severe accident manuals also had not planned on such a situation. Workers either could not find answers to their questions, or they found an answer that left them with even more questions, such as, "Where is this valve they say we need to open located?"

In addition, although the plant had several fire trucks on hand, none of the workers knew how to operate them. Fortunately, there was someone on location who did know how, although he was understandably reluctant because of the ongoing aftershocks.

At Fukushima, workers did their best, but without electricity and because of the damage to the infrastructure from the earthquake and tsunami, they could not overcome the rapidly escalating problems that eventually overwhelmed the plant.

No matter how well prepared a nuclear plant site believes it is, once an incident occurs they soon discover a few things they just flat didn't think of.

What's worse is discovering a few things they would have thought of if they had made a serious effort to consider and prepare for worst-case scenarios.

9: *Rumors Abound*

When an event happens, and nobody at first seems to know what's going on, the confusion is usually magnified when those who appear to be in charge make contradictory statements, which are then contradicted by some other expert.

By this time, everybody is an expert, at least as far as each confused and frightened individual is concerned, and because of this, nobody seems to question or see through some of the more incredible rumors that get passed around.

When people are already frightened, and nobody in charge is explaining anything, the people are literally ready to believe almost anything.

Three Mile Island

Three Mile Island certainly had its share of false or questionable news releases and rumors. At the time, nobody knew if they were going to be poisoned by radiation or blown up with the hydrogen bubble.

At any one time, several stories, some directly contradicting one another, circulated until Harold Denton took over as spokesman for all involved entities. Because he looked like an ordinary man-on-the-street and spoke to people in terms they could understand, they trusted him.

In the meantime, rumors went back and forth, some reported and some not, that a meltdown had occurred, that a meltdown had not occurred, that the hydrogen bubble had exploded and completely destroyed the town of Middletown, Pennsylvania, and that a plume of radioactive particles had been released and would contaminate most of the state of Pennsylvania.

Many people, not understanding radiation or having easy access to information as we do now with computers and search engines like Google, feared the plant was likely to explode like a nuclear bomb at any moment.

Anytime a situation like this develops, worst-case rumors develop right along with it.

Chernobyl

Chernobyl rumors were slow to get started because of the Communist Party's grip on news coverage. The method the Politburo planned to use was the usual "sweep-it-under-the-rug" plan, and to that end, not even the people in the nearby town of Pripyat, which existed to house the people who worked at the Chernobyl plant, knew the real status of the plant after the accident.

It is highly likely they would not have known of the accident at all if it hadn't been visible from many of the townspeople's apartment windows.

Once other countries began to detect the radiation loosed by the accident and demanded explanations, the Communist Party heads still kept a lid on information. They admitted only that an accident at a reactor had occurred, which was being mitigated, and the people affected were being given aid.

Naturally, a situation like this called for more explanation, and if the Communist Party would not give out more information, then the press would find other people nearby who would talk.

Indeed, they did, and consequently, many Western newspapers carried headlines about the 80 people killed at once and the 2,000 victims sent to hospitals. The *New York Post* topped them all, with a headline about the "mass grave" for the 15,000 victims of Chernobyl.

The most damaging rumor of all was likely the lie told by the Soviet leadership to the people of the town of Pripyat when the people were instructed to pack up for evacuation. The townspeople were told to take only what was needed for the next few days, when in reality, the leadership knew the people would never be allowed to return. Consequently, many Pripyat citizens soon found themselves in strange towns in dire circumstances, without money or clothing or the papers they needed to conduct almost any kind of business in the Soviet Union.

Once the town was evacuated and the government called in the workers known as the Liquidators to do the work of cleaning up the Chernobyl site and building the "sarcophagus" that would enclose the dangerous radioactivity, a new and even better rumor arose: Vodka was protective against radiation.

The Liquidators, in particular, discovered a lot of truth in this rumor. Although it didn't provide any protection against radiation, it did help their minds. The work they had to do was arduous and dangerous, and they welcomed anything that helped them through this time of service to their country.

Fukushima

The accident at the Fukushima Daiichi plant took a while to fully hit the consciousness of both Japan and the world. Due to the enormity of the natural disaster and the resulting deaths, the situation at the Fukushima plant did not become immediately apparent to everybody the way it would have if the earthquake and tsunami hadn't already attracted all the attention, worldwide.

It took the Japanese government a little time to realize just what was happening at Fukushima and what the results would likely be. It probably took TEPCO just about as long, because it likely bought into the idea that a bad accident could not happen, and a really bad accident was just not possible at a modern nuclear power plant.

Because of the severity of the natural disaster and the difficulty in communication caused by it, nobody could be sure of conditions at the plant itself. And at the plant, nobody was quite sure of the conditions because none of the instruments worked without power. All the workers knew was that if they didn't get some cooling water circulating through those reactors, bad things were sure to happen.

This state of affairs went on for several days. Nobody knew for sure what was happening at the plant, including TEPCO, the Japanese government, and the Nuclear Regulatory Commission in America. But scientists and nuclear specialists interviewed by television news programs were happy to speculate on what "could" happen. Worst case scenarios abounded, maybe even more than rumors.

In America, this opened the door for people with newly bought low-level radiation meters to rush to Pacific coast beaches to measure the radiation of the sand and post it on YouTube as proof that Fukushima radiation was already arriving.

Potassium iodide tablets sold out in America, as people prepared for the approaching "cloud of radiation" that would surely bring lots of radioactive iodine-131 to the entire North

American continent and contaminate American milk and food products.

The Japanese people struggled to get answers, such as where were any radiation plumes headed, and was it safe to send their children to school in their area? Japanese society was ordinarily very ordered and predictable, but suddenly chaos took over and no one stepped up to bring back order and a sense of control.

People were understandably outraged about the possibility of radioactivity in their areas, but one scientist had the answer: If you put on a happy face, the radiation will leave you alone. Present a smiling face, and you are in no danger from radiation.

Too bad nobody thought to tell this to the Soviet citizens involved in the cleanup of Chernobyl (although that may have been the real purpose of the vodka).

Later, this scientist said he had been struck by the sense of futility and depression among the people, and he only meant to encourage a more positive frame of mind. A positive attitude is known to help elevate the immune system.

We really do wish it was this simple!

10: The Big Question: Who's In Charge?

Probably the most important of all the lessons we've extracted from our studies of these three major nuclear accidents is that somebody must be in charge, and everyone must know where the proverbial buck stops. He or she must be in charge of everything, of answering to the press, of appearing on television, of briefing the President of the United States...this person must serve as the "face" of the accident. And we-the-people must view this person as someone who knows all the answers, including what's going on at any given time.

In short, we must trust him or her to be telling us the truth.

In every United-States-government-run scenario since the 9/11 terrorist attacks in 2001, the problem of *Who's In Charge?* has cropped up time and time again. This is an ongoing problem among government agencies, probably because each agency believes it should be Number One, and each agency's staff members tended to answer only to the head of their agency.

This is a huge problem, as we saw at Three Mile Island. Even though none of the staff was squabbling over the leadership position, nobody was clearly holding the position either. When the press arrived, looking for a spokesperson who could be counted on to have the latest information, they could not find one. Instead, they got conflicting information from the various entities involved.

The U.S. government has known about this problem since the late 1960s, when it caused horrific problems in the response to fighting wildfires in the American west. The Fire

Chiefs in Phoenix, Arizona, met in 1968 and formed the concept of sharing command among different agencies. It was honed during the early 1970s, when a series of massive wildfires damaged property and killed or injured many people.

Problems in managing an incident like an enormous wildfire often boiled down to communications problems and management problems rather than not having enough resources or using bad tactics. Each agency had its own management structure and procedures, which clashed when more than one agency tried to respond to the same incident. Each agency even had its own vocabulary, and problems ensued when a word had a different definition at each of the participating agencies.

This led to the creation of the Incident Command System (ICS) as we know it today. The ICS is intended to provide a framework for managing all aspects of an *incident*, where an incident is defined as an "unplanned situation necessitating a response." These are situations like natural disasters or man-made disasters, public health incidents like disease outbreaks, hazardous material spills, or terrorist attacks.

The key concepts involved in the Incident Command System are (1) Unity of command, in that each person participating in the operation answers to only one supervisor, (2) Common terminology, (3) Management by objective, (4) Flexible and modular organization and (5) Span of control.

Unity of Command, with each participant answering to only one supervisor, stops the problems caused by one person receiving conflicting orders from several supervisors and helps improve the flow of information. There is generally only one *Incident Commander*, who is in charge of the entire incident response. If the incident is a large one that involves several agencies, the incident commander is replaced by a

Unified Command, which consists of the head of each agency involved, each having equal authority.

Common terminology is intended to stop a lot of the confusion that can be caused when a word had a different meaning to each participating agency. ICS has a glossary of terms and uses defined terms to describe facilities, command roles and resources. This helps assure that everyone is on the same page when passing information along and giving and following orders.

Management by Objective means that an incident is "managed by aiming towards specific objectives," which are ranked by priority. The objectives are to be as specific as possible, attainable, and hopefully, achievable within a working time-frame.

A *Flexible and Modular Organization* denotes an organizational structure that can be easily expanded if necessary, and can be easily contracted as the incident resolves. Only those positions that are needed at the time are set up.

The *Span of Control* refers to the number of responsibilities and resources any one person can manage. Each person should manage between 3 and 7 people or resources, with 5 considered ideal. If a person is managing more than 7 people or resources, the structure needs to be expanded, and if one person is managing 2 or less, that position can usually be absorbed by another as the organization contracts.

The Incident Command System makes use of standardized forms that provide for accurate documentation of the incident in the form of Incident Briefings, Incident Objectives, Organization Assignment List, and many more.

All these things are designed to coordinate the activities of incident responders and to eliminate a lot of the confusion and turf warfare that developed during earlier incidents.

But these things are still a problem when more than one agency responds to an incident. Read descriptions of government-run scenarios designed to test response to nuclear incidents or biological warfare incidents, and you will soon realize that, although some things have improved, certain other things never change.

They still have huge problems determining the answer to: *Who's In Charge?*

Three Mile Island

The Three Mile Island incident occurred in 1979, well after the time of the Incident Command System, but it appears the system was never used there. Perhaps that is one reason nobody knew who was in charge.

There were several entities that could have been in charge. The company that owned the Three Mile Island plant, Metropolitan Edison, worked at downplaying the incident; the plant manager and operators were in a state of confusion trying to figure out what was happening; and the Nuclear Regulatory Commission, which was supposed to give wise advice to everybody in such situations, soon discovered it didn't know what was going on either.

Small wonder the news media felt lied to and stonewalled at every turn.

According to one of the NRC staff years later, he found himself involved in the situation as the plant operators held a pow-wow trying to decide what was happening and what to do about it. His opinion was considered as valuable as everyone else's, and he was struck by the fact that there appeared to be no clear-cut line of command inside the plant.

There seemed to be so many agencies involved, it is hardly surprising that the press had no idea which they should consult. They included the Pennsylvania governor's office, FEMA, the Pennsylvania Emergency Management Agency, the mayors of several towns located near the reactors, and of course, the Nuclear Regulatory Commission and the plant owner, Metropolitan Edison. The press was well on the way to developing a deep and abiding distrust of all of them before Harold Denton came on the scene.

Although he was not in charge, technically speaking, he had the knowledgeable but unassuming manner Americans like in leaders and he appeared to always know what was going on. He did an admirable job of serving as the "face" of the Three Mile Island incident.

Chernobyl

The hierarchy in the Soviet Union at the time of the Chernobyl accident was known. The question of who was in charge never came up because everybody knew the leaders of the Communist Party would make all the important decisions.

The only problem that arises in this type of structure is the one that we saw at Chernobyl: If the leaders of the Communist Party know nothing about what is happening, their decisions will likely reflect that, and they may not be the best decisions for that particular incident.

Fortunately, the Soviet Politburo consulted nuclear scientists, who managed to keep the Politburo from trying to keep the incident a secret. It appears to have taken some real work.

We can't help but wonder what would have happened if the Politburo had kept on denying what everybody in the world already knew. Very likely we would have read some headlines that made the "15,000 deaths" headline look paltry.

An incident needs to have a commander who knows something about the incident he is managing. If it's a wildfire, we need a commander who has been a firefighter. If it's a hazardous materials spill, we need a commander who has some knowledge of chemistry. If it's a nuclear reactor accident, we need a commander who knows something about nuclear power plants and how they operate.

And, we might add, he needs to be able to discuss these things in terms an everyday person will understand.

Is this too much to ask?

Fukushima

The accident at Fukushima greatly resembles Three Mile Island in terms of figuring out who's in charge.

Determining who was in charge was an iffy proposition from the beginning because of the way the Japanese bureaucracy had been set up in terms of nuclear power regulation. Several government agencies oversaw several different aspects of nuclear power plants, and the Prime Minister of Japan at the time saw himself as the head honcho of the whole mess. The chief executives of the Tokyo Electric Power Company would have likely claimed that position also, had they been in the country when the earthquake and tsunami struck.

The results were similar to what happened at Three Mile Island. Different leaders of different agencies knew different things, and reporters for the various Japanese media quickly became annoyed and mistrustful, especially when the chief government message was always, "Remain calm." That message became a sure sign to many that they'd better panic.

The problem with figuring out who was in charge was exacerbated by the natural disaster that had hit Japan. The earthquake had damaged a lot of infrastructure, which meant that communications were out in many areas. Even though there was a video connection between TEPCO headquarters in Tokyo and the Fukushima Daiichi plant, nobody knew what was happening, and the TEPCO people often didn't see fit to communicate with government officials what they did find out.

Then there was the problem at the plant with all the instruments that were no longer working, so the operators had no idea what was going on either.

The Japanese people finally got fed up and took over for themselves, much to the government's chagrin. They no longer cared who was in charge. So far as each household was concerned, the head of the house was in charge of the wellbeing of his or her family.

Fukushima is an example of a problem that could not possibly happen in a civilized, modern nation—until it did.

Who Should Be In Charge?

Each of these accidents would have been improved, if, from the very first, a spokesman had been appointed to serve as the "face" of the accident, somewhat like Harold Denton's role at Three Mile Island. This is no small thing, especially when it comes to comforting the people affected and to dealing with the press.

Harold Denton looked like nobody special. He wasn't spectacularly good-looking and probably wasn't a professional speaker of any sort. But he did know his subject, and he knew better than to cover up what was happening by employing terms nobody but nuclear physicists could understand.

He spoke to the Press like he would have spoken to his non-nuclear-educated friends, and he gave the impression that he was being straight with them. At no time did he give the impression that he was trying to hide certain information, or give an impression that was not quite the truth.

This may be too much to ask, because Harold Denton did not come to Three Mile Island to become the "face" of the accident. It just so happened that he was the right man who appeared at the right time, when there seemed to be nobody trustworthy around who knew anything.

When a serious nuclear incident of any kind happens again in America, we look for *Who's In Charge?* to be the major question on everybody's mind.

We hope FEMA and all the local responders who will be involved in the incident have learned this lesson well, and that they will choose, very carefully, the individual who will hold this position.

All we can say is, God help us if they institute a Unified Command (formed of the heads of participating agencies) that fails to designate that one trustworthy individual we-the-people can put our trust in to always tell us the truth, and not to hide anything from us.

If we don't know who's in charge, you can bet it will be déjà vu all over again!

11: *How The Next Event May Go Down*

After looking at incidents in the past, we can identify many similarities in how things developed. All the way from *It's Worse Than You Think* to *Who's In Charge?*, we have seen that the people involved in the incident seem to go through certain stages of denial and false belief, because they can't quite admit, even to themselves, that the worst has actually happened.

Much of this goes back to the major problem that affected all three of the events we have studied, namely the belief that nuclear power is so safe now, no accident can happen.

At Three Mile Island, this belief underlay the entire response to the accident. Nobody could quite believe that an accident had happened that could threaten much of the state of Pennsylvania, not to mention much of the East Coast.

At Chernobyl, this belief was practically engraved in stone. The Communist Party leaders did not want the citizens suspecting nuclear power might have a few drawbacks, so they actively promoted the belief that nuclear power was so safe, they could build a reactor in Red Square. When the accident occurred, their main concern was to keep a lid on what had happened and its consequences so the people and the rest of the world would not question the safety of Soviet nuclear power.

At Fukushima, every known redundancy in the modern world was in place. Even the backup diesel generators had backup battery-powered generators. They had battery-powered walkie-talkies and every other known safety device, including fire trucks on the facility grounds and protective clothing. With all that forethought, surely no accident could be so bad, it would overcome all those preparations!

But the accident that occurred involved an earthquake of greater magnitude than the plant had been built to withstand, and the tsunami that rolled in was far higher than what the plant's sea wall had been built to keep out.

Which all goes to show that the best-laid schemes of mice and men often go higgledy-piggledy, especially when those mice and men get cocky and think something like, "Even God Himself couldn't sink this ship."

Government Is Slow To React

When something does happen, the government faces the same problem everybody else faces at first—finding out just what the heck is going on. As you have seen, this is not easy to do, especially when nobody else knows what's going on either.

Once the government does find out what's going on, the same old problem rears up...who's in charge?

At Three Mile Island, we still aren't sure, except that Harold Denton made an excellent stand-in for whoever should have been in charge.

At Chernobyl, the Soviet Politburo was in charge, and it got busy right away trying to bury the whole problem. The area around Chernobyl was evacuated far later than it should have been, because the government did not want to panic the people.

At Fukushima, we still aren't sure who was in charge, although the Japanese Prime Minister certainly appeared to believe he was. He did succeed in slowing down some of the response efforts, along with the government agencies that were supposed to oversee nuclear power plants and the officials of TEPCO.

That's what we must remember about governments. They seem to exist to slow things down. Since government tends to be a collection of committees, all these committee meetings have to be held before a consensus opinion is arrived at as to what to do. By that time, the crisis has often moved on to an entirely new and different stage.

The purpose of the Incident Command System in the United States is to avoid a lot of this conflict and enable a timely response to the incident. It remains to be seen whether or not the system will work when a real nuclear incident occurs in America.

With the things we've seen in mind, let us now take a look at how the next nuclear event in America may go down.

A Government Scenario In 2011

As you may imagine, the government is also interested in this same question, so periodically FEMA will run a "scenario" with the participation of various emergency management groups, to see what the likely outcome will be, and hopefully to identify any shortcomings.

One of the more recent scenarios was done by Lawrence Livermore National Laboratory, in conjunction with Sandia National Laboratories and many other emergency response agencies in 2011. The scenario involves an improvised nuclear device exploded by terrorists in downtown Washington, D.C.

The publication, *National Capital Region Key Response Planning Factors for the Aftermath of Nuclear Terrorism*, carefully noted the likely blast effects, fallout effects, radiation effects and even the agriculture embargoes likely to result. They also noted the importance of having messages to the public prepared beforehand, and people trusted by the citizens to deliver the messages.

From all we have seen, however, most of the real-life accidents produce some weird developments that nobody could have expected, often having to do with the reactions of those same citizens.

The government, for instance, seems to think citizens will shelter in place as instructed, when many of them will actually hop in their vehicles and flee. A nuclear incident anywhere in America is going to panic many people. Since Americans as a whole are very mobile people, look for a goodly portion of the population of a targeted area to simply leave, whether it's safe or not. They won't care what the authorities say. They just want out of there.

As for our list of "lessons," look for them to be repeated in some form or fashion.

It's Worse Than You Think

When the incident first begins, you may hear about it on the news or read it on the Internet. At first, it won't sound like much. Even if the news source hypes the event, you will likely discount most of what is said.

So will the people involved in the incident.

Nobody Knows What's Going On

As the incident develops, if it is nowhere near you, you will likely notice that follow-up news stories don't seem to add very much new information. That is likely because the news people can't find out what's happening, because the people involved won't know for sure what's happening.

Operators Disbelieve Their Instruments

Depending upon the incident, we look for this to repeat in several ways. If it's a nuclear power plant accident, the operators will likely not believe some instrument that is telling them the reactor core is in danger. If it's a nuclear explosion anywhere in America, many first responders will likely not believe the high readings they get for fallout in certain areas.

First Responders Are Ill-Equipped

In America, we have some of the best-equipped first responders ever in terms of *equipment.* In a nuclear incident of any kind, they will be decked out with respirators, protective suits and the latest and smallest radiological meters ever invented.

The problems we are likely to see with first responders in a nuclear incident in America will occur because of two reasons:

One: We have allowed a new system of radiation units to be taught and used in America, one which is exactly 100 units different from the old system based on Roentgens that has been in use here since World War II.

Two: We have allowed a certain group of scientists to perpetuate the idea that there is "no safe level" of radiation, in spite of all the evidence to the contrary. This is known as the *Linear No Threshold Theory*, meaning that there is no "safe" level of radiation, below which no damage to the human body occurs.

For this reason, many first responders are all too likely to be terrified of radiation, believing that if they receive any dose at all, no matter how small, it will result in cancer. Believe it or not, there are many scientists out there who worked in the national laboratories during the nuclear testing programs of the 1950s through the 1970s who never developed cancer in spite of receiving various doses of radiation in their work.

Compounding this fear will be the confusion inherent in using a second system of recording radiation units that is exactly 100 units different from the old system. One Sievert (the new system) is equal to 100 Roentgens (the old, familiar system). Since many modern radiation meters will read in either system, we look for a lot of confusion with responders switching between the two systems on their meters, with maybe even some deaths resulting. *Stress generates confusion and degrades performance*, and responders will suffer enormous stress during a nuclear event in America.

Imagine taking a reading of 25 on your radiation meter, and believing you are using the Roentgen scale. On that scale, 25 Roentgens is a relatively safe dose of radiation, resulting in no symptoms of radiation sickness. However, if the scale had been inadvertently switched to the Sievert system, a reading of 25 is plenty of reason to worry. This is a reading equivalent to 2,500 Roentgens, a dose resulting in Level III radiation sickness and death.

So we look for some first responders to accidentally switch scales when taking readings, and for others to compare their Roentgen-scale readings to their Sievert-scale readings, which will compound the confusion already running wild. We expect a few to refuse to go into areas of radiation to perform rescue operations, even if the levels of radiation are fairly low.

Also, another problem with first responders we may see is the same problem military recruiters are complaining about: So many young Americans today are not physically fit enough to serve their country. Many are on drugs for everything from mental problems to diabetes and high blood pressure. This isn't even considering the effects of bad diet and lack of exercise.

When a nuclear incident happens in America, look for all these things to add to the atmosphere of general confusion that will surround the incident. You may even see the 180 RLH maneuver (turn 180 degrees and run like hell!).

Rumors Abound!

Any time something like a nuclear event happens in America, the robust American news media gets busy on it. With the ease of Internet research these days, any reporter can quickly gather any information needed, unlike the days of Three Mile Island, when reporters had trouble researching what went on in a nuclear power plant and how one operated.

We also look for various "experts" to be consulted, who will further muddy the waters. There are many nuclear experts today who should know better, but who will convince those who listen to them that any little escaped ray of radiation is going to cause cancer or Acute Radiation Syndrome in any person unlucky enough to be struck by it. We will probably see entire neighborhoods evacuated over ridiculously low radiation levels.

We also might expect some really interesting rumors to arise, although we have no idea what they might be. Just be aware that everything you hear during an incident is not necessarily true. Wait until the rumor is confirmed by more than one trustworthy source before you accept that it's likely true.

Who's In Charge?

In spite of all the work done on developing the Incident Command System and practicing on terrorist scenarios, we still look for the question of *Who's In Charge?* to play a major role in the initial confusion surrounding the next nuclear incident in America.

We may be told right off who's in charge, only to have it become obvious very soon that this person is really not in charge, because everyone is listening to someone else. Or worse, the same situation that occurred at Three Mile Island will arise again, with the news media unable to figure out which of the "experts" is telling the truth.

It seems that in every government-run scenario since the 9/11 terrorist attacks, *Who's In Charge?* was a big stumbling block almost every time. In spite of the ICS system that was designed to mitigate those problems, responders tended to answer to their own agency head rather than to the person assigned the duty, especially if that person was from a different agency.

The problems with the shared command of the Unified Command were much the same. People tended to answer to the head of their own agencies, in spite of the concept of "equally shared command," which sounds really great until they try to implement it.

No matter how well-researched and carefully structured a concept is, people will still be people.

A Nuclear Plant Accident In America

One morning you may wake up to the news that a "minor accident" has occurred at the nuclear power plant near you.

As the day progresses and you tune in for regular updates on the incident, you may notice that the whole thing seems to be growing worse. Not that you're any expert, but you quickly pick up on the fact that nobody seems to know for sure what's going on in spite of all the happy-face statements about a *stable situation* and *no threat of a meltdown*.

Remember that the powers that be don't want us to realize, first-hand, that nuclear power can be quite dangerous if certain things happen. Until the situation becomes clear, authorities will try and soft-peddle events, but once the news media gets wind of the real story, you'll likely experience information overload.

This is the time to make sure you are prepared to shelter-in-place. Any time there is a nuclear power plant situation, if the situation should go really bad, emergency planners are supposed to have already prepared evacuation plans for the areas within 10 miles of the plant in all directions.

Assuming you live just outside this area, or in an area down-wind of the plant, it is possible you may be threatened by a radioactive plume, should the plant really go haywire.

Usually, you will be safest if you plan to remain in your own home rather than try to flee the area, whether an evacuation is called for or not. All too often, evacuations cause worse trouble in that traffic is jammed up and doesn't move very far for hours on end. This means you may be sitting in your car, unmoving, when the plume arrives, and you will receive more exposure to radiation than you would have if you'd stayed in your home.

Or you may flee in the direction you are told to go, only to discover that the plume has changed its mind and decided to go that way also. Recall the troubles with the Japanese attempts to evacuate villages out of harm's way, only to discover the people were moved to an area of even higher radiation than the one they'd left. The same thing happened at Chernobyl, but events there did not become clear until years later.

Unless you are close to the plant and in the direct path of a plume on the order of those from Chernobyl, your best plan is to take shelter inside your own home.

The only way to know for sure is to monitor the news carefully and to take radiation readings with your own meter. The government will be taking readings in areas all around the endangered power plant, but the only way to know for certain what the readings are at your home is to take your own readings. The homes one block from yours may have much heavier or much lighter readings, depending on the wind.

For help in choosing a radiological meter, see Volume I of Dr. "B"s Radiation Series, *How To Choose A Civil Defense Radiological Instrument: Geiger Counters & Dosimeters*. If you live near a nuclear power plant, **you would be wise to plan your actions well in advance and prepare for the worst**.

You may not know for a while what the best course of action will be. Judging from the past, you cannot go wrong sheltering in place until it becomes clear which areas received the heaviest dose of radioactive particulates. By that time, the authorities will likely have picked the best evacuation routes, if you should need to evacuate.

If you live within ten miles of a nuclear power plant plant, you may wish to plan ahead for a time of sheltering in place if the radioactive plume should be heavy. For advice on how to create a fallout shelter within your own home, see Volume II of Dr. "B"s Radiation Series, *Your Home Fallout Shelter: How To Ensure Your Family's Health & Survival In A Nuclear Event*.

A Nuclear Bomb Explosion In America

If a nuclear bomb should be exploded anywhere in America, our lives will never be the same again. Panic will ensue, even in the tiniest towns and rural areas, much as it did after the 9/11 attacks.

If the United States is at war with a country that is armed with nuclear weapons and a nuclear strike occurs, look for more to occur. That is the time that you definitely want to be prepared to shelter in your own home, with your own radiological monitoring equipment. Multiple fallout clouds can arrive over a period of days to weeks if we have a nuclear war.

This would be a worst-case scenario indeed, and if this should occur, know well that our country is no longer prepared with public fallout shelters stocked with two weeks of provisions the way it was prior to the *Mutually Assured Destruction* era of the late 1970s until now. **You will be on your own, and it is up to you to protect your family**.

If you believe that this might occur in the near future, then study every book in Dr. "B"s Radiation Series that pertains to your situation. Once the first strike occurs, America will essentially be frozen in place and it will be too late to make any preparations. You will be on your own with what you have on hand at that moment.

If a terrorist group manages to explode an *improvised nuclear device* (IND), most likely in a big city, that is a different matter. The damage will be far less than what would occur with the large nuclear bombs that will be used in warfare.

If you live near the stricken city, your preparations will be similar to those you would make if the nuclear power plant near you blew its stack. An IND explosion, although it is a nuclear bomb, would still be much smaller than the nuclear bombs used in modern warfare.

Nuclear fallout will be your main concern if your home is located far enough away from the blast to avoid damage, but close enough to receive fallout from the prevailing winds. A large nuclear bomb may send radioactive dirt and debris high into the atmosphere, where it can be carried long distances by some of the higher-level winds.

Even an improvised nuclear device may send debris high into the upper atmosphere, where the winds may carry fallout to distant places.

Models have indicated that fallout from an improvised nuclear device of about 1 kiloton would deposit dangerous levels of fallout from the point of the blast to an area about nine miles away in the direction of the prevailing winds, within minutes of the blast. This is the area, if your home should be located there, that will benefit the most from preplanning and being able to take quick action.

If you live within 10 miles of the blast, or you live in an area where the radioactive plume is expected to arrive, your best bet is to plan on sheltering in place. Stay inside and do not try to flee the area. Chances are you would wind up driving right into the most radioactive areas, or being caught in your car as fallout comes down.

For this scenario, you definitely want your own home fallout shelter, as described in our book, *Your Home Fallout Shelter*. Levels of radiation from fallout in this area can reach 200 to 300 Roentgens, which can cause you or members of your family to develop Level II radiation sickness if you remain in it for any length of time. Your own home fallout shelter will help you implement the known strategies for dealing with radiation: *Time*, *Distance*, and *Shielding*.

How Will You Know Fallout Is Coming Your Way?

There are excellent computerized prediction programs these days for plotting the probable path of a nuclear plume, but the wind sometimes has other ideas. Wind currents do not always go in the directions they usually travel. If a plume of radioactive fallout shifts course within a few minutes of the blast, how will you know it is coming your way unless the emergency authorities notify you?

That's the problem. You probably won't know fallout is coming. Unless you live far enough away that authorities have time to run the programs and estimate directions of travel, you may have little to no warning, especially if you live within 10 miles of the blast area. Plume prediction models would not work fast enough to warn anyone within that 10-mile radius, and that's assuming emergency authorities could get their act together fast enough to run the models.

If you hear, on the news or from a concerned friend, that a nuclear device has been exploded in the city near you, you should immediately get out your radiological meter and begin monitoring radiation levels. If you have a low-level meter and you know the usual background levels of radiation in your area, you can easily detect if and when fallout begins arriving. A high-level meter will tell you this also, and will also be able to keep tabs on the levels as they rise above the upper limits of a low-level meter's detection capabilities.

It may be that you are working outside and do not know a nuclear device has been exploded in a nearby city. If you should suddenly see fine particles like sand, dirt or ash dusting the area around you, do not remain outside in it. Go inside, get your radiological meter, and check it for radioactivity.

If the fallout is radioactive, immediately get into the shower and remove it from your skin and hair. Radioactive fallout that is left on the skin or scalp for several hours can cause beta burns, as the beta-emitting radioactive particles decay.

Also get your pets inside and bathe them, as fallout that remains on their coats can cause burns. If you have pets and are concerned about them in a nuclear event, see Book IV of our Radiation Series, *How To Protect Your Pets, Livestock and Home Garden In A Nuclear Event*.

Stay Put!

Once fallout begins coming down in your area, do not try to pack up and leave. You will only get more exposure to radiation, and you may well wind up stuck in traffic with highly radioactive fallout building up on your vehicle roof.

Fallout that is heavy enough to cause injury is also going to arrive too quickly for you to outrun, and that is assuming you know which direction to run in. Get yourself, your family members and your pets inside your house, where the roof and walls provide some shielding and distance from the fallout.

This is where your preparations for a home fallout shelter will come in handy if you studied and prepared in advance.

You and any members of your family present should immediately put your preparations into action and get inside your home fallout shelter.

Bring your radio inside with you, and your meters. No matter what is told you about the levels of radiation in your area, it can be heavier or lighter in various sections of the locale. You want to know what it is where you and your family are located. You also want to make sure your shelter has no leaks.

Fallout that comes down immediately after a nuclear blast is usually composed of the heaviest and most radioactive particles. Some of these particles decay very rapidly, within the first day after arrival. As the most radioactive of the isotopes decay, radiation levels begin to fall fairly rapidly. If you have a meter, you will be able to monitor the decline of the radioactivity around your home. Most experts estimate a stay of 1 to 2 weeks in a fallout shelter, depending upon the size of the blast and whether there were multiple bombs set off.

Chances are, if fairly heavy radioactive fallout has fallen in your area, you will be notified of an evacuation. You will be told when to leave, where to go (usually to an emergency shelter) and which route is the safest to take. If you decide to leave, you should turn off all ventilation systems (air conditioners or heaters) in your home and lock all the windows and doors.

In you evacuate in your car, keep the windows closed and the ventilation system turned off until you reach a safe area. Take your meters with you and keep track of the radiation levels as you go.

If you go to an emergency shelter, remain there until authorities decide radiation levels have decayed enough to allow residents to return. Generally, this may take anywhere from one week to two weeks, and even then, it is always possible emergency officials will decide the area must undergo a "cleanup" before the residents return.

Should You Evacuate?

This is a good question. Many people who have made preparations for a home fallout shelter will not see any reason to leave, even if authorities insist.

Many definitely do not wish to go to a government shelter of any kind. The noise and chaos on top of everything else are too much for many to handle. You could always evacuate yourself to a motel a hundred miles away, but that can be costly, especially for any length of time. Plus, it is always possible the wind can change direction and cause your motel to be closed down also.

If you have meters and have made proper preparations, and your meters show that your home fallout shelter is doing a good job of protecting you from radiation, you may not wish to take the chance of going through radiation in order to hopefully arrive at a safer place.

This is your decision to make. If you feel your shelter is not sufficient to protect your family, then by all means leave and go to a safer area.

However, if you have prepared, remain in your home fallout shelter. Authorities may go from house to house in order to roust out all the residents, but if you remain silent and keep all pets quiet, they are likely to assume you already left the area. You should also be prepared to protect your premises and your family, because looters are a possibility in spite of the radiation.

Aftermath

Once the radiation levels have dropped, or the cleanup has been done, whether from a nuclear plant accident or an atomic bomb blast, you should be allowed to return to your home. Or you can emerge from your home fallout shelter and behave as if you have just arrived home from visiting your relatives out-of-state for the duration of the crisis.

As life slowly returns to the new normal, where very little will be the same as the old normal, you will begin to look around and hear the tales your neighbors have to tell of their evacuation experiences. This is when you will either realize how much better off you were because you remained in your own home, or if you evacuated also, you will realize that you would have been much better off if you had built your own home fallout shelter.

The moment these thoughts hit your mind, know well that it is time to begin preparing for the next nuclear incident in America.

We know only one thing for sure: Sooner or later, there will be another!

Radiation Safety Limits

0 to 50 Roentgens	Considered Safe
50 to 200 Roentgens	Level I Radiation Sickness
200 to 450 Roentgens	Level II Radiation Sickness
450 to 600 Roentgens	Level III Radiation Sickness

Level I Radiation Sickness: Less than 5% deaths. From 5% to 30% of exposed people may develop acute symptoms of nausea and vomiting within 4 hours of exposure. A temporary reduction in blood platelets and white blood cells may occur.

Level II Radiation Sickness: Less than 50% deaths. From 60% to 75% of exposed people may develop acute symptoms nausea and vomiting within 4 hours of exposure. Severe blood changes, hemorrhage, and hair loss.

Level III Radiation Sickness: Greater than 50% deaths. One-hundred percent of exposed persons will develop acute symptoms of nausea and vomiting within 4 hours of exposure. Severe blood damage, hemorrhage, and hair loss, with up to 80% deaths in less than 2 months.

Generally speaking, the higher the dose of radiation received, the faster the onset of symptoms. At Chernobyl, the doses received were so high, firefighters noted a metallic taste in their mouths and a severe headache before they developed nausea and vomiting that forced them to stop their work.

In a couple of "criticality accidents" in experimental laboratories (cases where an unexpected event caused a brief, high release of radiation), the radiation-blasted person immediately felt as if he was burning and tingling all over and developed almost instant nausea and profuse vomiting.

As has been said by many experts in the field, if the radiation dose is high enough for you to actually feel, it's also high enough to kill you.

Here is a small chart you can copy onto a 3 x 5 card and keep with your meter. In an actual incident, stress tends to degrade your memory, and radiation is something you do not want to take chances with.

1 Roentgen = 1 Rem = 1 Rad = 0.01 Gray = 0.01 Sievert

0 – 30,000 CPM = Safe

Greater than 30,000 CPM = Questionable

```
1R = 1Rem = 1Rad = 0.01Gy (Gray) = 0.01 Sv (Sievert)
0 - 30,000 CPM = Safe
Greater Than (>) 30,000 CPM = Questionable
```

Radiation Limits For Outside Activities

Less than 0.5 R/hr: Essential precautions should be observed, such as respirator mask, goggles, wide-brimmed hat, full-length raincoat down to your ankles, boots with their tops up under the bottom of the raincoat, whiskbroom and gloves. You can only stay out in this for a few hours per day. *You should sleep in the shelter.* You must give your exposed body time for healing and repairing!

0.5 R/hr to 2 R/hr: Same precautions as above. Activities should be restricted to essential duties, such as put out a fire; secure food and water; acquire medical supplies; save a life. *Do not stay out more than two hours per any one day*!

2 – 10 R/hr: Hold time outside the shelter for no more than a few minutes a day for absolutely essential purposes!

Greater than 100 R/hr: *No outside activities permitted*!

The above precautions were designed for people staying in fallout shelters, so they would have an idea of how much work they can do outside the shelter if the necessity should arise during the time they are staying inside their home fallout shelters.

The main thing to remember in a situation where you have to go outside the shelter during a time of fallout is that *your body must be given the chance to repair the damage from radiation.* This is why you can work outside several hours day when radiation is less than 0.5 R/hr, but *you still must sleep inside your shelter.*

Link To Downloads Page

For a list of downloadable files concerning radiation:

http://www.chemicalbiological.net/Downloads.HTML

For a FEMA publication concerning the scenario of a terrorist improvised nuclear device (IND) exploded in downtown Washington, D.C.:

http://www.chemicalbiological.net/NCR%20Key%20Response%20Planning%20Factors.pdf

Resources

Actions For Survival: Saving Lives in the Immediate Hours After Release of Radioactive or Other Toxic Agents, Allen Brodsky, 2011

A Field Guide to Radiation, 2012, Wayne Biddle

Chernobyl 01:23:40: The Incredible True Story of The World's Worst Nuclear Disaster, 2016: Andrew Leatherbarrow

Demand for Personal Geiger Counters Soars In Japan, Kawase, C, Reuters, May 25, 2011

Effects of a 10-kt IND Detonation on Human Health and the Area Health Care System, Institute of Medicine (US) Committee on Medical Preparedness to Respond to a Terrorist Nuclear Event, Workshop Report, Washington (DC): National Academies Press (US), NCBI Bookshelf, 2009

Fukushima Daiichi Nuclear Disaster, https://en.wikipedia.org/, 2018

Fukishima: The Story of a Nuclear Disaster, Lochbaum D, Lyman E, Stranahan SQ, Union of Concerned Scientists, 2014

Introduction to Radiological Monitoring: A Programmed Home Study Course, HS-3, 1972: Staff College, Defense Civil Preparedness Agency

Japan Acknowledges First Radiation Death Among Fukushima Workers, Tsukimori O, Sheldrick A, Macfie N, Reuters, September 5, 2018

National Capital Region Key Response Planning Factors for the Aftermath of Nuclear Terrorism, Buddenmeier BR; Valentine JE; Millage KK; Brandt, LD; Published by FEMA & The Department of Homeland Security, November, 2011

***NOVA**, Preparing for Nuclear Terrorism*, http://www.pbs.org/wgbh/nova/military/preparing-for-nuclear-terrorism.html, February 24, 2003

Public Protection From Nuclear, Chemical, and Biological Terrorism, Health Physics Society 2004 Summer School, Brodsky, Allen; Johnson, Raymond H., Jr.; Goans, Ronald E, Editors, 2004

Radiation Safety In Shelters, Federal Emergency Management Agency, 1983

Radiological Defense; Textbook, 1963, United States Department of Defense, Office of Civil Defense

Shelter in Place, https://en.wikipedia.org/, June 2018

Stay Tuned to Learn How to Evacuate, https://emergency.cdc.gov/radiation/evacuation.asp, Centers for Disease Control and Prevention, October, 2014

The Effects of Nuclear Weapons, 3rd Edition, 1977, Glasstone S, Dolan, PJ, U.S. Department of Defense and Energy Research & Development Administration

Three Mile Island: A Nuclear Crisis In Historical Perspective, J. Samuel Walker, 2004

Three Mile Island: The Hour-By-Hour Account of What Really Happened, Mark Stephens, 1980

Under-Exposed: What If Radiation Is Actually Good for You? Ed Hiserodt, 2005

About the Author

Dr. Charles S. Brocato is a scientist and author who has written extensively in the fields of surviving chemical and biological warfare and nuclear warfare. He is also a longtime martial arts instructor, weapons/firearms instructor, and a map & compass instructor, specializing in *how to stay alive*. He is also a nutritionist, a counselor in the field of nutrition and an award-winning French chef.

He holds degrees in biology and mathematics with advanced studies in the areas of biochemistry, microbiology, and advanced mathematics, and a D.D. (Doctor of Divinity). He has also completed numerous studies in areas of Emergency Management with FEMA and MetEd.

Two of his previous books are available on Amazon in print form: *The Two-Fold Chastisement: Visions of the Coming Earth Changes*, and *Chemical/Biological WarFare: How You Can Survive*.

This book is the sixth in a series of Radiation titles he has planned in the area of surviving a nuclear attack or incident.

Write him at: **csbauthor@chemicalbiological.net**

Other Books By This Author

Dr. "B"s Radiation Series

Book I: *How To Choose A Civil Defense Radiological Instrument: Geiger Counters & Dosimeters*

Book II: *Your Home Fallout Shelter: How To Ensure Your Family's Health & Survival In A Nuclear Incident*

Book III: *Nutrients Against Radiation Damage & Injury In A Nuclear Event: Something You Can Do Before It's Too Late*

Book IV: *How To Protect Your Pets, Livestock and Home Garden In A Nuclear Event*

Book V: *The Dirty Bomb: Should You Worry?*

Book VI: *How A Nuclear Incident In America Will Probably Go Down: What We Learned From Three Mile Island, Chernobyl and Fukishima*

Other Topics

Rosemary: The Healing Herb of St. Martin de Porres

The Two-Fold Chastisement: Visions of the Coming Earth Changes

Chemical/Biological WarFare: How You Can Survive

Christian Meditation: Understanding What Meditation Is

Dr. "B"s Motto:
"You Can Never Have Too many Meters!"

www.ingramcontent.com/pod-product-compliance
Lightning Source LLC
Chambersburg PA
CBHW020435220526
45464CB00002B/707